The Price of Peace

The Price of Peace

*The Future of Defense Industry
and High Technology
In a Post-Cold War World*

William H. Gregory

LEXINGTON BOOKS
An Imprint of Macmillan, Inc.
NEW YORK

Maxwell Macmillan Canada
TORONTO

Maxwell Macmillan International
NEW YORK   OXFORD   SINGAPORE   SYDNEY

Library of Congress Cataloging-in-Publication Data

Gregory, William H.
    The price of peace : the future of the defense industry and high
technology in a post-cold war world / by William H. Gregory.
      p. cm.
    ISBN 0–669–27950–1
    1. Defense industries—United States. 2. High technology
industries—United States. 3. Industry and state—United States.
I. Title.
HD9743.U6G7 1992
338.4′ 76233′ 0973—dc20   92-29050
                                              CIP

Lexington Books
An Imprint of Macmillan, Inc.
866 Third Avenue, New York, N.Y. 10022

Maxwell Macmillan Canada, Inc.
1200 Eglinton Avenue East
Suite 200
Don Mills, Ontario M3C 3N1

Macmillan, Inc. is part of the Maxwell Communication
Group of Companies.

Printed in the United States of America

printing number
1 2 3 4 5 6 7 8 9 10

To my wife, Gini, whose encouragement
helped to make this book possible

Contents

Preface

Unexpected peace obviously will reshape a defense establishment sized, equipped, and technologically geared to wage four decades of Cold War. Not so obvious is what peace will do to an economy that prospered on the military dollars that created high technology and the jobs and equipment to go with it. The influence of peace on our industrial economy is the subject of this book. While it centers on the 1990s and the evolution that peace will bring, its origins are in pell-mell demobilization after World War II and in the defense cuts that were made as this country disengaged from Vietnam.

Both were part of my life. I was one of a generation of Americans who reached abrupt maturity in uniform during World War II, in my case as a naval aviator. As the war continued, we thought we might spend the rest of our possibly short lives there. Then Japan collapsed, and we found ourselves abruptly in pin-stripe suits again and not sure how to face a world where people no longer shot at or dropped bombs on each other.

For us the transition was less traumatic than it was for those who joined the military later, for we became part of an unprecedented wave of American economic expansion. Yet my generation did not sever all connection with the military. For many of us it was an intimate connection, spanning the entire Cold War. As an editorial manager, and a writer on the business and economics of aerospace for a leading magazine, *Aviation Week and Space Technology,* the development of a whole range of mind-boggling military aircraft, missiles, and electronics was daily grist. So it happened that in the late 1960s and early 1970s, I watched from the front row as a darker side of this story unfolded. First came an unpopular undeclared war and the fiscal and monetary mistakes made in financing it. These unleashed double-digit inflation that compounded the ordeal of the subsequent disengagement from Vietnam. Thousands of engineers and production workers were laid off in what was a defense and technological recession.

Watching the interplay of defense research and development with nondefense technology and the economy over 40 years, it struck me that if the Cold War should end one day, the effects on U.S. industrial life would be profound. My experience convinced me that defense stimulated technology in many ways, and thus economic activity, and I was convinced its withdrawal would be the economic equivalent of a smoker quitting cold turkey. Without national support for the development of technology and a strategy for dealing with a curtailed military role, today's consequences could be more far-reaching than those of the post-Vietnam decade. The Cold War has ended, and this book is an attempt to measure its ramifications on America's technological and industrial future.

Hundreds of discussions and meetings with engineers, scientists, and military officers over 40 years influenced this book; their names are too numerous to list individually. I want to thank them collectively for their ideas and to thank those who are quoted individually here for their invaluable insights. Specifically, too, I want to thank my editor at Lexington Books, Beth Anderson, for her helpful comments on the manuscript and especially her patience in bringing it into coherent form.

William H. Gregory

1

That Elusive Peace Dividend

Protocol that May Day in 1991 called for flying the flags of all the North Atlantic Treaty nations on the speakers' platform. They snapped in a typical hot, gusty desert wind—incongruous under a typically inert, cloudless, cerulean blue sky—that sprayed dust on the distinguished visitors and on 300 local watchers. Soviet and U.S. flags flew to one side as a backdrop for the torpedo-shaped gray aluminum shell in the foreground, an unprecedented diplomatic pairing only a couple of years earlier. Military marching band music boomed from loudspeakers. Davis-Monthan Air Force Base in Arizona—in a fitting touch, the place where the U.S. mothballs its out-of-fashion military aircraft and missiles—doesn't have much in the way of live music.

Speeches began, a bit stilted because of the halts for English-Russian and then Russian-English translations. Then a five-man crew in white coveralls walked to the six diamond-blade, bright-orange power saws on each side of the shell. The first saw balked, then started. A second did the same. Again, symbolic of a retreating military way of life, the four others were spares.

Two designated point men slipped blue protective face shields in place and began to cut the metal skin at each end of the cylinder, longitudinally toward the middle where they met. Cut saw engines, ran the military script. Bring up forklift and pick up cylinder. With a stab at military flourish, the crew tugged at the cylinder, and abruptly it split into symmetrical halves. At that the crowd applauded spontaneously. Later, a few spectators sidled up to the shells and tried to wrench off pieces of metal for souvenirs.

So there was a garden party air, with wives and children bubbling and laughing and cheering. Be that as it may, it was also historic.

The power saws had chewed through the last of U.S. Tomahawk ground-launched cruise missiles destroyed over three years of mutual inspections by agreement with the Soviets in the INF treaty, the intermediate-range nuclear forces olive branch clasped by the two principal Cold War adversaries in a manner unprecedented since World War II. In retrospect, the INF treaty was the breakup of the ice that had frozen national strategies of the United States and the Soviets for almost 40 years.

U.S. Army Pershing 2 ballistic missiles also fell to the saw blade, 846 missiles in all of both types based in Europe to meet a Soviet nuclear missile threat. A week after the cruise-missile ceremony, the last one was destroyed in Texas. The Soviets, under the eyes of a similar U.S. team, destroyed 1,846 missiles, mostly the medium range SS-20s that were targeted at Western Europe.

General Lieutenant Vladimir I. Medvedev, the head of the Soviet nuclear risk reduction center that monitored the U.S. weapons destruction, made a brief speech at the Arizona gala, alluding to the coincidence with May Day in the socialist holiday hierarchy, a day once marked by massive military equipment parades in Moscow. Medvedev, not in uniform like his American military counterparts but instead in a standard-issue, loose-fitting Soviet business suit, hailed this May Day as the embodiment of the first practical step toward eliminating nuclear armaments. Since that May Day ceremonial, events tumbled one on top of the other faster than citizens could comprehend what they meant.

Soviet withdrawal from Eastern Europe. Accelerating defense budget cuts in the United States and its European allies. A failed Soviet putsch, an eclipse of the Communist party that drove the Cold War, then disintegration of the Soviet monolith itself into a shaky commonwealth. A unilateral beginning of a standdown of the centerpiece of U.S. grand strategy for dealing with its Cold War adversary: massive nuclear forces ready to launch in minutes of detection of a hostile strike.

Arms control still rests on verification and compliance as well as good will. Military adventurism has not been permanently repealed. Yet the need of the fragmenting Soviet economy for surcease from incessant technomilitary competition with the United States left little realistic expectation that the former Soviet empire could mount a serious military effort. Even a shooting war in the Persian

Gulf had its own lesson in its amazing brevity. Who anywhere in the globe can threaten this country's military position? No sooner had the Cold War with the Soviets ended than the hot political war began over how large a military establishment the United States should maintain.

So began the dismantling of the mighty military structure the United States had built since the Soviets closed off Berlin in 1948. With that dismantling comes another fundamental and perhaps more far-reaching change: the start of a standdown of the equally mighty military and defense industrial research and development effort that was the underpinning of those forces. As the power saw motors died in the desert, as the crowd disbursed with smiles for a drink at the base officer's club, the country was about to find out in real time, not theory, what does happen as defense sinks back to a smaller factor in U.S. industrial operation, perhaps to a level not seen since war in Europe threatened in the late 1930s.

The Day After the Cold War Ended

America's federal government throughout the Cold War was awash in war plans, an understandable preoccupation. Of peace plans, it had few. Yet one of the most profound implications of peace is what its effect will be on the economy, in both the macro sense and the micro, on communities that lose bases, defense plants, and jobs; on technology; on research and development; on manufacturing. None of these should be ignored, but the one to watch especially is the path of military research and development.

In the 1980s, the defense share of all federal research was as high as 68 percent (in 1988). Japan, with treaty limits on its defense posture, spends only 3 percent of its government research money on defense. Preunification West Germany, also subject to treaty restrictions on its military, spent 12 percent. Defense research spending has propelled the U.S. economy, with early development of the computer, commercial aircraft navigation systems, and cellular telephones among its by-products, although how much the economy has benefited is controversial. Unfortunately, the financial figures are misleading. Most defense research and development goes for production startup, testing, validation, evaluation, and the like.

Defense basic research, that which is not tied to specific applications, amounts to about $1 billion annually out of a $40 billion total 1992 fiscal-year Pentagon research and development line item.[1] Beyond that is a definition question of how much other exploratory research, which also has potential payoff for the rest of the economy, adds to generic military research. Estimates run from $4 to $8 billion. Because of the difficult-to-measure commercial fallout from defense, whether the United States is outspending or underspending its international industrial rivals is unclear. Holding its basic research constant is the Pentagon's plan. But hard times bring bad news, and good intentions tend to fall victim to expediency. With U.S. business and industrial privately funded research under pressure owing to economic retrenchment, a defense research drop could mean more bad news for U.S. technology.

Consider these other elements of defense on the U.S. economy (Table 1–1), as cited in a Congressional Budget Office study:[2]

These states all have broad-based economies, the study points out, so the effect of state output is not as large as in less industrialized states like Alaska, Hawaii, and Mississippi, where defense is more than 8 percent of the total economy. And New Mexico, South Carolina, Washington, Virginia, and Maryland will lose more than 1 percent of their output, based on defense cuts planned in 1991,

TABLE 1–1
Defense Spending in Selected States—1991
(billions of dollars)

State	Direct and indirect defense spending
California	$99.3
Texas	42.8
Virginia	38.4
New York	25.5
Florida	25.2
Pennsylvania	22.0
Ohio	19.8
Georgia	17.7
Maryland	17.2
Washington	16.7
Massachusetts	16.1
Missouri	14.9

before reductions accelerated. Many communities fear larger cuts. A regional study group in Los Angeles estimated that the $9 billion in defense contracts that flowed into the Los Angeles area in 1990 could fall, at the low end of its forecast, to $3.5 billion by 1995. A drop of that magnitude, the task force study said, could cost up to 400,000 jobs—a direct defense and ripple effect—and as much as $84 billion in personal income. While those are extremes, the forecasts show how severely defense shrinkage could touch broader economies. Every state will lose something, even if only a little. Farm states like North Dakota and Montana could face base closings as the Air Force Strategic Air Command shrinks. Thus there are implications for national economic stability as the inevitable cuts broaden.

Perhaps more important in the perspective of national policy is people. Defense has absorbed a large share of this country's technical talent. Defense industry chiefs and university researchers agree that about one-third of U.S. engineers and scientists worked in defense in the 1980s. Diversion of technical manpower into defense work was a necessary sacrifice in times of national danger but was onerous for commercial industry and universities, and onerous for the national welfare as well, considering the loss to the economy as a whole of the brightest engineers and scientists, the loss in new ideas and products, and the loss of innovation in commercial industry.

Well might the question be asked why commercial industry and universities failed to compete for talent, but more than handsome salaries for defense work were involved. Engineers like technical challenge, and defense is—or was—a more exciting arena for a designer than analyzing muffler brackets. As defense shrinks, concerns like these will reverse. To see how, recall the 1950s and the conventional wisdom then. Soviet engineers had just launched the first Sputnik and developed intercontinental ballistic missiles with hydrogen warheads. More engineers were coming out of Soviet than American training schools and more of them were working in defense. It seemed to American commentators then that too many of ours were designing new washing machines, refrigerators, cars, or, in a note with a hollow ring now, television sets. As defense shrinks, so will the market for engineers and worry will return about whether the United States will have enough technical talent to compete in the post–Cold War industrial game.

Whether defense research and development or weapons production helped or hurt the U.S. economy was an argument that continued through much of the Cold War. Defense gave us the jet airliner and the electronic computer. It contributed in various important ways to the development of the transistor and the subsequent evolution of the semiconductor. Though some critics recognize that these technologies grew out of defense, they argue that the transfer slowed down with more specialized and sophisticated weapons in the later stages of the Cold War. But the late Richard D. DeLauer, director of defense research and engineering in the Reagan administration, listed these later defense-spawned technologies: computer time sharing, packet switching for faster communications transmission and fifth-generation computing, encompassing natural programming language, artificial intelligence and its related expert systems that seek to put the experience of a skilled and experienced worker into a computer's memory.

As the Cold War threat changed with time, another kind of worry arose. Conceding that facing down the Soviet nuclear threat was important, government, academic, and industry commentators suspected that the United States was carrying a disproportionate share of the burden. Now that question becomes more pointed. does the post–Cold War threat, the country wonders, warrant the kind of leadership in defense spending that has characterized the last half-century? Former Prime Minister Margaret Thatcher startled Washington, or at least the media, with a speech there during the wave of euphoria just after victory in the Persian Gulf. Only America has the strength to defend the free world, she asserted. A tribute to be sure, but also a sobering thought for Americans. Is she right that only America can man the future ramparts, that only America can lay a mantle of peace over a turbulent world with its military preeminence?

While Mrs. Thatcher may have been intuitive about America's strength, there is another aspect to consider: whether that open handedness with defense will be acceptable to the American public given the abrupt changes inside the Soviet Union. A commitment of that kind of public spending, fluctuating between 23 percent and 28 percent of federal spending in the 1980s, along with a commitment of one-third the nation's technical best to defense, will require reappraisal. Putting this talent back into the civilian world is attractive. Doing so is something else. Peace is going to come at a price, just as the Cold War victory did. While the United States carried a large share of the cost of victory, its allies, especially Japan, learned better how to apply the fruits of basic research to commercial technology than did America in the latter half of the century.

Such disconcerting developments have not escaped notice in Congress. "Other industrialized nations—particularly in Western Europe and Japan—construct their technology efforts with a greater emphasis on economic development over military development than the United States does," the Congressional Office of Technology Assessment noted.[3]

> They are increasingly demanding that military technology support commercial development wherever possible. In Japan, almost all technology is developed for commercial purposes, and some of it is then exploited for military uses. What is appropriate for these other nations is not necessarily good for the United States, since neither Japan nor any Western European nation aspires to be a superpower. However, these are nations with which the United States is competing economically. We may be able to benefit from making both military and civilian R&D do double duty.

At least some in industry and the Pentagon might look askance at the notion that none of America's international competitors hunger for world leadership. Some see Western Europe 1992 as one massive power bloc, Japanese-dominated Asia as another, and the United States left to make the best of what it can work out in a trading partnership with Canada, Mexico, and perhaps South America. Others wonder whether, with the Soviet Union disorganized, broke, and no longer a common enemy, historic rivalries with Germany and Japan will reignite. A few suspect the next U.S. war will be with Japan. Probably that's emotion and frustration, for Japan's puny military force would have trouble dealing with Saddam Hussein, let alone with a first-class military power. To forge a revived Japanese military force would take years or decades. Besides, Japan has learned it is far less costly in money and people to wage economic, not military, war.

Where the Congressional report was on target was its stress on setting double duty out of military research and development. This country has indeed invested huge sums and immense manpower in military weapons innovation. Common sense dictates two things:

- Military development in the next century must utilize commercial technology far better than it has.
- Commercial technology must benefit from the nation's investment in military technology.

Nothing Lasts Forever

While the Cold War dominated national strategy, a much subtler kind of conflict had indeed broken out. With its World War II enemies decimated, and its European allies devastated as well, this country luxuriated in a plant and equipment infrastructure unbombed and unshelled when Germany and Japan capitulated. While rebuilding Europe with the Marshall Plan, America also dominated the postwar global economy unchallenged. Industry grew fat and complacent in a couple of decades after the war. By the 1970s, though, a little bad news began to seep into this happy scene. Germany and Japan rebuilt, Europe revived, Asia caught a new vitality. One American industry after another lost its global leadership. Worse, they were losing share in a heretofore private preserve: the domestic U.S. market.

America's decline in the international marketplace, in the reliability of its industrial product, was compelling enough to stir interest in whether the government ought to do something about its vacillating approach toward industrial technology and industrial competitiveness in general. Though defined differently by different advocates, as those who equate it with protectionist trade practices, there were calls for a U.S. technology policy or an industrial policy as the competitive aspect of U.S. industry magnified in the 1980s. "Technology policy?" some scoffed. "That's a nonsubject." Others snorted: "We don't have any."

Both skeptics are right, in a way, allegorically. In fact, this country does have a technology policy, or rather, over the years, it has had a lot of science and technology and industrial policies. Even its deliberate attempts at not having a policy is a policy, a policy to do nothing. Looked at another way, America's support of a large defense and space research and development effort, both in industry and in an array of government laboratories, is a de facto technology policy—or perhaps technoindustrial policy. Although not necessarily intended as a technology or industrial policy, this has been a richly funded effort for four decades—more than $500 billion—and one marked by the lack of integration that might be expected with goals that were geopolitical rather than scientific or industrial.

Semantics play a bigger part in the question than they should. Inside the black wrought-iron fence that guards the White House staff from the real world, the triumvirate of then chief-of-staff John Sununu, budget director Richard Darman, and Michael Boskin, chairman of the Council of Economic Advisers, had a reputation as the iron triangle, the fortress against industrial policy. Sununu—"Say No No" to advocates inside the Washington beltway for a more coherent technoindustrial policy—is a Massachusetts Institute of Technology–trained engineer who ought to understand the technology game. Yet starting with David Stockman in the Reagan administration battle over helping out the U.S. automobile industry against Japanese import competition, U.S. industrial policy was synonymous with bailing out declining industries. Defined that way, industrial policy is an obvious nonstarter in Washington.

Now Sununu is out. His abrasive dealing with Congress and his perceived role in running domestic policy at a time of recession cost him his job late in 1991. With or without Sununu, industrial policy is not popular among Republicans, who see it as bailouts, but more popular among Democrats, who see a bigger role for government in job creation. Again, how industrial policy is defined is crucial. As industrial bailing out, it is one thing. As federal technology policy to foster U.S. industrial competitiveness while defense technical influence shrinks, it is another. Or perhaps it is picking technical winners or losers, another industrial policy lightning rod. Despite failures like the abortive U.S. attempt to build a successful commercial supersonic airliner in the late 1960s, government has been picking winners in military technology for at least a century. Japan, Korea, Singapore, and much of Western Europe approach industrial development and technology differently, and, certainly in Japan's case, with results to show for it.

Congress recognized the problem as Japan's trade surplus with the United States rose. The House Committee on Science, Space, and Technology formed a Technology Policy Task Force and undertook to study the issue. Its agenda, laid down in 1987, observed in more formal language the bald truth about U.S. technology.

America does not have a formal and articulated policy for stimulating and guiding technology development," it said. We do have numerous laws as well as many ingrained patterns and practices that govern the way that we envision, develop and utilize technology. These guidelines and influences, viewed collectively, form an "ad hoc" national policy for technology.

Given the sometimes stumbling performance of U.S. industry in domestic and international competition, the Task Force made it plain that it was time to take a fresh look at the ad hoc governance of technology. Much useful comment emerged in the Task Force report, and in hearings and reports of other congressional committees and public and private commissions. But no semblance of consensus has emerged.

What Happens Next?

As for the industrial and economic side of defense, it began to shrink as a factor in the U.S. economy in the latter part of the 1980s as the Reagan defense buildup ran out of favor in Congress. Democracy in the Soviet Union and then dismemberment accelerated the downward pressure.

At one extreme, defense could compress from a big industry and the military from a big employer to their relative levels in the 1930s, on the eve of World War II. In the defense contracting world, this possibility is core meltdown. While some in Congress call for deeper cuts, others prepare to fight specific reductions, such as trimming the National Guard and reserves.

Initially, with Soviet collapse, the Bush administration offered a plan to cut defense roughly 25 percent by the middle of the 1990s. Pentagon leaders have intimated that this would not have a major impact on the economy. Defense in the United States amounted to 5.5 percent of gross domestic product in 1990, which Donald J. Atwood, deputy secretary of Defense, compares with 25 to 30 percent in the Soviet Union. Cutting back defense outlays between 1991 and 1997 in the administration's now-superseded 1991 plan—including half a million uniformed personnel plus at least 85,000 civilians and the National Guard and Reserves—will amount to just under 2 percent of gross national product (GNP). Echoing the Administration's "not to worry" attitude, Atwood added: "I think there still will be great opportunities for people still in the defense business and in growing commercial operations." In the severe 1981–1982 recession, for a benchmark, U.S. GNP declined 2.8 percent.

Before the end of 1991, the Pentagon began to face reality. First, the 25 percent cut was in constant dollars. So Atwood candidly pointed out that the defense budget in the Pentagon's original plan might be only slightly smaller by 1995 in 1995 dollars and crucially dependent on the inflation factors used by the Pentagon—in other words, not much of a peace dividend. An illusionary peace dividend will outrage Congress. Just to be prepared, the Pentagon began to look at other options: cutting 300,000 more uniformed personnel. Army divisions could fall from 18 to 10 instead of the 12 surviving in the early plan; Navy carriers from 13 to nine instead of 12; and Air Force tactical fighter wings to fewer than 26.5, including Guard and Reserve units expected to make up 42 percent of the force by 1995. To put that in perspective, 26 fighter wings is what the Air Force fought for in the stringency of the 1970s, and it developed the less expensive F-16 as an alternate to the more expensive F-15.

By early 1992, George Bush recognized that even 25 percent was not feasible. He raised his five-year shrinkage to 30 percent, but then drew a line and said, no more. That could be a futile gesture, too, to a Congress and public hungry for money to put into other programs. But the peace dividend after the Vietnam War shrinkage of the military was part illusion, and it will be so this time, too. In his fiscal 1993 budget, President Bush proposed $16.8 billion worth of reductions in defense. Just the concomitant increase in social welfare and entitlements, housing, and education amounted to $48.9 billion. Rising interest on the estimated $4 trillion 1992 gross national debt—interest whose level before deducting federal interest income was close to the size of the defense budget in 1991—will add more costs without any new social or civil infrastructure programs. While the federal government will dress up its books with defense cuts, the economic impact of losing high-paid jobs and high-technology plants on national and local tax revenues and in higher unemployment costs will have to be netted out of the defense budget savings. So much for the peace bonanza.

Not that surgery on the defense budget is necessarily bad news either for the military or for contractors, though they may not see it that way. A friend of mine in one big defense conglomerate told me in frustration that the kind of money allocated for defense made sloppy controls or outright fraud almost impossible to stop. Besides that, both contractors and military slip into bad habits: too many support people, too many engineers, too much business by rote rather than by innovation. Dieting could help the health of our national security. Even more telling: those F-15s, F-16s, F-18s, and cruise missiles that carried Desert Storm into Iraq had their genesis in end-of-Vietnam, tight military budgets of the early 1970s. Pentagon acquisition paperwork, specifications, standards, and unnecessary testing add needless cost to military buying.[4]

A similar thought was expressed in the Congressional Budget Office study[5] as the fiscal 1993 budget went to Capitol Hill. Reducing the federal deficit with defense savings, it concluded, would stimulate the economy to absorb military drawdowns by increasing national savings, lowering interest rates, raising domestic investment, and reducing borrowing overseas. Such was the intent of the Omnibus Budget Reconciliation Act of 1990 that came out of the White House budget summit with Congress, channeling any peace dividend into deficit reduction rather than new federal discretionary spending, defense or otherwise. In the two years after the budget summit, congressional Democrats periodically attacked that act, proposing to move billions in defense dollars over to social programs or Soviet relief. All attempts failed. But defense dollars for deficit reduction is a precarious proposition as politicians look for new revenue for new spending. Besides, deficit reduction may be deflationary, and not stimulative but depressing.

Two kinds of models were used in the congressional study. One assumed that defense cuts would be offset quickly by lower interest rates, increased investment, and an improved trade balance, that is, that markets anticipate changes in fiscal policy. The other model did not. Both forecasts, using the original administration reductions, were in the range of a half to three-quarters of a percentage point drop in real gross national product in 1993 and 1994. But the model assuming markets anticipating changes brought recovery from the bottom faster, two years earlier than the less optimistic one. Using larger reductions, approximating the new Bush budget plan, the drop in gross national product ranges from three-quarters of a percentage point to over a full point. Employment changes bottom out at around four-tenths of a percentage point in both models, using the larger and more realistic reductions, and take about six years to recover to 1992 levels.

All studies like this one are dependent on certain assumptions, principally of all things being equal. If the rest of the economy is stable, recovery from the defense downsizing will occur in five or six years, and to be manageable, forecasts have to assume that. Yet things are not equal either in the short run or in the long run. For instance, how fast and how well solutions are found for the structural and competitiveness problems of the rest of U.S. industry are unknown factors in the depth and duration of the defense downturn. Similarly, as the study says, "the link between defense spending and economic prosperity has long been a subject of dispute." One proposition is that defense is a drag on the economy, though, the study concludes, mainstream economists consider defense neither a drag nor a stimulus. Still another view is that high defense spending is crucial to prosperity because of capitalism's tendency to underconsumption and stagnation, and that technical talent working on defense would be unemployed otherwise.

Empirically, the latter view has a point. Anyone who grew up in the depression and drought of the 1930s knows intuitively how federal deficit spending, or at least just federal spending, can stimulate the economy. Federal funds and, believe it or not, federal wisdom saved many a family farm until the rains came back, a couple of years before World War II. By that time military spending, building up for World War II, revived industrial prosperity that Herbert Hoover had predicted was just around the corner in 1930. Hiring the unemployed to rake leaves vanished as rearming the country brought factory jobs instead. A relative of mine went from an underpaid shop-teaching job to foundry manager of a busy aircraft-engine plant.

When defense contracts were cancelled wholesale after World War II, fears of a return to the Great Depression revived. So much pent-up demand for cars, appliances, and all the other civilian products whose production was suspended for the duration of the war instead turned into a new boom. Then came the Berlin block-ade and the Cold War. Except for the Vietnam standdown of the 1970s, defense spending continued at hitherto unprecedented peacetime rates, and for the most part recessions were short com-pared to the Great Depression of the 1930s. Though there was a decade of slump as the country cleansed itself of Vietnam, Ronald Reagan came along. Though hardly intending to emulate the Roo-sevelt revival of the 1930s, he unleashed a new deluge of deficit and defense stimulation and the "go-go" decade of the 1980s. An unprecedented tax cut ensued, to be sure, but neither Reagan nor Congress really quit spending. Sympathetically, consumers and commerce went on a borrowing binge, too, and the economy launched into hyperspace. Now, in the 1990s, the borrowing hang-over is here and the largest defense drawdowns since World War II are coming.

Defense winddown? "I think it is more than a winddown," says Ray Waddoups of Motorola, whose Government Electronics Group's survival depends on how the transition works out. "We are probably in the last months of the death of this industry as we knew it in the 1950s, 1960s, 1970s, and 1980s. In the 1970s when we cut back, all we did was cut back. Then we came back exactly the same way we had always done business." Not this time. Waddoups believes: "We've got to change permanently how we do our busi-ness, even though it's not well defined yet how." Some things are clear in his mind though: coming is more employee participation, bottom-up, the Japanese style; going are more layers of manage-ment and more of the people in them. Defense may still command whole plants or whole companies run exclusively to labyrinthine regulations and specifications, but fewer of them. Defense will use more commercially available components. Custom designing and building will be unaffordable. Paradoxically, pure-defense surviv-ors, with fewer competitors, will be strong. The Cold War floor that preserved the defense industry mold after Korea and after Viet-nam is gone. Now it's free fall.

Winding down defense does not have to be a problem, in theory. Past record suggests that it will become one: the 1970s all over again, but worse. Then the combined military budget procurement and research and development spending that fueled the defense in-dustry dropped about 25 percent between the 1969 peak and 1974 trough fiscal years. True, much of the decline was in consumables, the beans and bullets of a shooting war, but the principle that discretionary items get hit first was demonstrated, for the total defense budget fell only about 6 percent from peak to trough as inflation boosted payroll and operating costs like fuel. That downturn was enough to worry big contractors. Scores of small suppliers dropped out of defense or out of business altogether. Overseas competition has intensified since, and as specialty manufacturers see their defense markets invaded and shrinking, there will be intensifying cries for protectionism.

Now the reduction in the total defense budget (in real terms) will be more severe. A National Science Foundation board member, Arden L. Bement, Jr., of TRW, whose defense interests are large enough for him to worry, believes that the reduction will fall between the original Pentagon plan of a 25 percent cut across the board by the mid-1990s and the call by Congress and the public for as much as a 50 percent cut. "There is only a certain rate at which you can really draw down," he said, "because of long-term commitments—and also the pain. Congress isn't going to take the rap for community impacts and all that sort of thing."

This is not just another of the cyclical downturns in defense that occurred about every seven or eight years during the Cold War. Defense shrinkage will be so large that it will signal the end of the country's ad hoc defense technoindustrial policy of the Cold War. At the same time, U.S. industrial competitiveness is still suffering from the challenge that arose in the 1970s. Like one stream flowing into another, these two trends are going to demand unceremoniously that the United States decide what, if anything, its future industrial and technology policies will be as it builds its new strategic posture in national defense and commercial, technological competitiveness.

2

Restructuring the Military

Soviet collapse left both our national strategy and military structure under new attack—of introspection, of how to go about drastic change. With the world rolling upside down in a matter of months, everything—policy, organization, strategy—went up for grabs, big or small. Army divisions began to melt away, Navy battle groups came under budget fire, and, of all the services, the Air Force began to reorganize most fundamentally. "Reorganization" is the right word, for as Air Force Secretary Donald Rice told Congress in early 1992: "We are not paring down the Air Force. We are building a new, smaller Air Force from the ground up."

Most fundamental was the change in U.S. strategic nuclear forces. Symbolizing the nuclear showdown with the Soviets was the postwar creation of the Air Force Strategic Air Command, the command and nerve center of America's nuclear retaliatory missile and bomber arsenal. At the height of the Cold War, bomber crews stood day-long alerts, ready to fly in minutes in case of attack. Specially equipped flying command posts were airborne around the clock.

In little more than a year of precipitous peace, the constant watch began to evaporate. By the summer of 1991, fewer nuclear armed bombers stood cocked alerts with nuclear weapons on board. As the Soviet threat faded, SAC's bomber force dwindled from 700 to about 200, itself a sign of less tension and less money for the military. SAC's B-2 stealth bomber program truncated; the fleet will stay small. SAC's Looking Glass command posts, KC-135 cargo-tankers fitted with electronic consoles and always with a general officer on board, once airborne 24 hours a day, went to earthbound alert. Then, in the fall, President Bush pulled Strategic Air Command bombers off alert entirely, and some intercontinental missiles as well. This included withdrawal of U.S. tactical nuclear weapons from Army and Navy forces. When arms control agreements are settled with the remnants of the Soviet Union, the Air Force will stand down all its newest Peacekeeper intercontinental missiles and cut the number of nuclear warheads on its only ICBM on duty, the Minuteman III, from three to one. Gone, then, will be the much-feared MIRV multiple independently targeted warheads.

During the Cold War, SAC's troop strength swelled to well over 200,000 and its auditorium-sized underground electronic command post south of Omaha was sealed off by beret-wearing special troops, attack dogs, and bank-vault doors. As part of a peace-spawned Defense Management Review, a blueprint agreement with Congress to whittle down U.S. force strength, and the Pentagon's later Base Force Concept SAC will shrink about a third in personnel—to about 85,000—in line with the rest of the U.S. military apparatus.

In an odd coincidence, General Curtis LeMay, who built the Strategic Air Command into a premier command at the beginning of the Cold War, was a hero to SAC's first post–Cold War commander, General Lee Butler, as an Air Force academy cadet 30 years ago. Now General Butler has the job of building down SAC in the Cold War's last act. Few military commanders want a legacy as a force cutter, but General Butler tells it like it is. "Our corporate culture has been strategic nuclear deterrence for 40 years, and properly so because we faced a rather daunting and for many years overly hostile Soviet threat," he told an interviewer soon after taking his new job. "But companies that don't respond to those changes in their environment go under. One day they just dry up. . . ."[1]

Instead of flying airborne alerts, SAC generals began preaching to Chambers of Commerce or service clubs anywhere that the command, and its $850 million per copy—or more—B-2 bomber, have a conventional war role. General Butler is talking about a second triad, beyond land-based missiles, submarine-based missiles, and bomber nuclear delivery systems. These are reconnaissance aircraft, tankers, and conventionally armed bombers, beamed at what the Pentagon views as the most like future military threat: regional, Third World–country wars like the one in the Persian Gulf.

Even more drastic changes are looming as the Air Force plans to eliminate SAC as a major command and move toward a dual Air Combat and Air Mobility Command structure. Despite the shock to the Air Force, SAC, as of June 1, 1992, officially became the Strategic Command, a nuclear-force planning rather than operating organization infiltrated by a growing contingent of Navy ballistic-missile submariners. Reorganization's thesis nevertheless is a good one: cut back the notoriously large support tier in the U.S. military and get colonels flying airplanes again instead of sitting at desks. That will be slow, for the Air Force—along with the other ser-vices—has been busy with rank escalation. Deescalation will be at-tempted, eliminating most deputy or vice commander jobs in the Air Force at both senior and junior officer levels. Adversity is gen-erating much-needed changes: more responsibility and indepen-dence for field commanders, along with less administration and more flying for them.

On top of that, the Air Force is combining its two military buying commands. These, the Air Force Systems Command, which carries out research, development, testing, and initial production of equip-ment, and the Air Force Logistics Command, which handles their supply and support in service, will become the Air Force Materiel Command. There is a message in consolidation: the endless search for the ultimate technological Holy Grail is out of fashion, and coming into fashion are cost control, fewer personnel and com-mand layers, and even more attention to reliability and mainte-nance in weapons requirements. In the meantime, there is a process that is part evolutionary and part horse trading in how the military will thin its ranks.

A Maze of Tradeoffs

Tradeoffs were exactly what the meeting was all about in a hotel banquet room in Raleigh as the woodsy North Carolina piedmont steamed in unseasonable heat and humidity in the spring of 1991. Fifty of the hundred-odd invited participants were green suiters, far more stars on epaulets than eagles or oak leaves. The rest were mostly business gray suiters. A core of CEOs, presidents and vice presidents, salted with a few sport coats from academia, were a bit deferential—and a bit resentful—to the stars and the smaller sprinkling of political appointees and senior civil servants from the Pentagon. Among them: retired General Maxwell Thurman, who commanded George Bush's invasion of Panama that preceded Desert Storm; Stephen K. Conver, who, as assistant secretary for research, development, and acquisition, the Army's weapons buying chief, conceived the meeting; Dr. John Foster, the head of the Defense Science Board and 20 years previously a respected director of defense research and engineering in the Pentagon, and Norman R. Augustine, a former Army chief of research and development who now heads Martin Marietta Corp. The room itself, with dark and subdued wood paneling but not gloomy, seemed to fit. For it was an august group with a problem on its collective mind.

Several of the Army generals there were lately back from Saudi Arabia and points north, from Desert Shield and Desert Storm. They were the good news that lightened the otherwise sober day. They had war stories to tell that were, as war stories go, unusually full of successes and low casualties—for the friendlies, that is. They were testimony that the much maligned high-tech stuff built for them by the industrialists in the room has confounded the armchair critics by winning in days.

All this was the obverse of the problem that the meeting had been called to consider. As the organizers of the meeting made clear, the Army was not going to be buying much more in the way of smart weapons, at least for a few years. Layoffs by defense contractors were underway before George Bush decided to go to war in the Middle East. Naturally, defense spending rose a bit with the war, but did not revise the budget summit caps on spending in the deal the administration struck with Congress. The inexorable downward trend in defense spending, off 22 percent from a peak in 1985 through 1991, is headed down further. Military planners find this disturbing, for they know how fast things change in an unstable world. Yet they don't have much to point to as a prospective military threat to the United States. A monolithic USSR was partitioned overnight and its new states are near collapse, the Warsaw Pact Eastern Bloc military alliance disappeared quietly; the bumbling campaign of Saddam Hussein in the Gulf War left a plain enough message that there isn't much around militarily to take on the vast U.S. defense apparatus built up over four decades after a spasmodic demobilization after World War II.

As the off-the-record message of the meeting made clear, this was a post–Cold War world. Not only was the Army running out of procurement money, but also there was no alternative in sight. Why this was so in an enormous defense budget of almost $300 billion takes understanding of how the Pentagon budget system funds the forces. When the president cut his budget summit deal with Congress that led to the Omnibus Budget Reconciliation Act of 1990 and to the Budget Enforcement Act that fenced off the peace dividend, dropping defense ceilings, later lowered, to $295.8 billion in fiscal year 1992 and $292.5 the following fiscal year, the Pentagon went through yet another, but far larger, of its incessant reviews—the Defense Management Review. To stay within what the Pentagon calls the defense topline, that is, the grand total of personnel, operations, procurement, and research and development accounts, it was necessary to cut uniformed personnel. At Raleigh, the Army was in the first wrenching adjustment to shaving force levels, a more technical name for uniformed people, from about 780,000 at the start of the decade to 535,000 by the end of the five-year planning period in 1994. In retrospect, that may be too optimistic. Similar cuts will come to the other uniformed services. In other words, a cut of about one-third in uniformed strength. Not that people shrinkage came overnight; the Air Force started in 1985 and had cut 123,000 from its uniformed forces by 1992 and planned to cut 55,000 more, to 428,000, by the end of the 1995 fiscal year.

Beyond that will be civilian layoffs, a rare RIF, a reduction in force, in the civil service if retirements and attrition don't thin ranks enough. Then will come layoffs of defense workers in private industry. As one senior general asked both rhetorically and incredulously, did his cohorts recognize that the defense establishment was going to dump close to a million people paid about $30 billion annually onto the economy, many needing to find new work, in the next four or five years? Add in contractor work force shrinkage, and the total approaches twice that at least. Was anyone, he wondered, thinking about the consequences, on the people themselves and on the nation at large? Those aren't net job losses, for some will find new ones and some will retire. But the 1990–1991 recession, to give scale to the defense reductions, saw 1.6 million jobs disappear. While the services dropped a million uniformed people after Vietnam, that was essentially cessation of the draft. This time career people are facing the axe.

Congress and the Pentagon are indeed wrestling with the consequences of the first cut, what to do for the uniformed people themselves. One manifestation: a pull-no-punches Army advertisement that asked whether America's workforce was bankrupt and suggested the answer was yes. It went on to offer help to industry in placing former soldiers through its Army Career and Alumni Program database and 800 telephone number.[2] The military has always expected a sizable chunk of its enlistees to depart after their first tour, usually four years. A tougher decision for both soldier, or airman, or seaman, and a retention battle for the service, comes after tour two. By then, the enlistee has put in probably eight years and is accelerating into the prime of his experience level as far as the military is concerned. Yet he has at least 12 more years to go to qualify for a pension and a shot at a second career. His military pay may not be buying enough housing or schooling for his family, and a decade seems a long wait for a better-paying civilian job. Yet it is with this group, those who committed after a second tour to a military career, that the layoffs will start. To avoid bouncing uniformed people outright, Congress approved, but did not initially fund, alternatives of an annuity or a lump-sum payment as incentives for voluntary separation. No question, though, the military fears it is going to ram some of its people.

Oddly enough, the Army found out that when it forecast its future people costs through the early 1990s, a one-third cut in forces was not showing any savings. The manpower account stayed stubbornly the same, reductions offset by projections of the effects of inflation and rising military pay and fringe benefits, done paradoxically in the name of retention of career people.

Manpower is one of the big accounts into which military budgets are divided. Operations and maintenance is another, the money for field exercise training, bases, gas and oil, food, maintenance of tanks and helicopters—all the things it takes to keep a force in the field. The Army agonized over that account, but the general reaction was that operations money equaled readiness. No one was in any hurry to tamper with money that meant an effective military force.

Still another big account, though not as big as the first two, is research and development. One thing none of the services wants to do, besides risking combat readiness, is to tamper with the genesis of future weapons. Those preeminent weapons in Kuwait were conceived not in the open-handed Reagan era, but a decade before—a measure of how long the weapons cycle is from concept to deployment. So, in the name of keeping the technological base strong, the Army decision was to hold at least the fundamental research part of research and development constant, something the other services are doing as well. That left procurement as the last of the big accounts, the money that pays for production of existing weapons. Given the ceiling of the defense topline budget, something had to go, and procurement was it.

"What can you do," asked a senior Army civilian, "given that kind of topline?" To try to stave off squeezing weapons production, the Army argued with civilian financial analysts in the staff offices of the Secretary of Defense. The response was: "The reason you guys don't get any procurement money is because you're protecting the force structure." Perhaps that may have been true in the past. But the Army argues now that when it reduces its force structure by a third and still can't afford money for procurement, that criticism is specious. Turning a bit emotional, the Army civilian snapped: "We think we've made a case that we need more money for procurement."

His goal runs counter to one of the many laws coined by Martin Marietta chairman Norman R. Augustine when he ran Army research and development two decades ago: if the total defense budget is down by 20 percent, procurement will decrease twice that much. When defense procurement goes down, it hurts the economy, for procurement money goes to big contractors who pay it to second and third subcontractors, suppliers, and job shops, all of whom create well-paying jobs, usually in manufacturing or research. In the 1993 budget projections, Augustine's ratio held. The Department of Defense topline in outlays in 1989 at $295.6 billion will slip to $275 billion in 1997, or 7 percent. Procurement spending at $81.6 billion in 1989 will decrease to $61.3 billion, down 25 percent, or three times the percentage decline of the total defense budget. To put all this into perspective, these numbers were generated before the president backed down from his original 25 percent defense cut proposal. So they may be academic. Compounding the confusion was the alternation between current dollars and constant dollars in the budget discussions. Reductions in current dollars, without adjusting for inflation, look much smaller in percentage terms than the promises of politicians made them sound. In effect, budgets are holding down future growth, rather than cutting back. Again, so much for peace dividends.

As justified as its case for more procurement money may seem to the Army, the chances for the Secretary of Defense, the White House, or Congress to relent on the defense topline before the middle of the decade—or even beyond—are negligible. Certainly the defense budget will go down more than the caps agreed to in the White House budget summit with Congress, a summit that preceded complete Soviet collapse. Reality penetrated the White House and the Pentagon when George Bush proposed in his 1992 State of the Union message to drop defense 30 percent, not the original one-quarter. Usually the way the game is played, Congress will go for 40 percent next; the administration will draw a new line in the sand at 35%. Then Congress will ask for 50 percent; the administration will draw a new line at 40 percent. Surprisingly, in the ensuing fiscal 1993 budget jousting, Congress, sensing the danger of more job losses in a recession, beat back several attempts to take more out of defense than the administration proposed, to find more money for social programs or even for aid to the new Soviet states. Capitol Hill assaults on defense funding will be incessant, nevertheless. Congress for its part wants "sound" defense budgets: deep cuts, all in someone else's district.

Salvaging Weapons Programs

Down the ring-shaped corridors of the Pentagon from the Army sits the Navy in much the same box over the future of naval aviation. Its programs for new aircraft to carry it into the new century are in tatters. A precursor came in the fall of 1990, one the Navy brought about itself when it cancelled a new patrol aircraft on the drawing boards, the P-7A that was intended to replace its fleet of P-3 anti-submarine surveillance aircraft. Patrol aircraft don't get the priority or attention of fighters or bombers that attack the enemy. So the original P-3 was an inherently less-expensive adaptation of a commercially designed aircraft, the Lockheed Electra transport sold to the airlines in the late 1950s. In winning the P-7 contract, Lockheed agreed to keep the kind of parts and design commonality with the

P-3 that the P-3 had derived from the Electra. No more than a year into the detailed engineering, Lockheed realized that it could not do so. Lockheed tried to raise the price of the airplane to cover the higher costs it expected. Contractor and service got close, but never together. Finally the Navy cancelled the program, but for default on the part of Lockheed rather than the more typical convenience of the government. Because of the default, Lockheed was left to sue for, and if it lost, to eat the extra money it had spent to try to meet the contract specifications—about $300 million.

Cancellation of the P-7 preceded peace with the Russians, but, for two reasons, it was a bellwether for metamorphosis of weapons buying. Not only did it establish the principle that the Pentagon would now force erring contractors to swallow their losses rather than allot more money, but, more important, it also established that the services would not insist on having new hardware if costs got out of hand.

No sooner had the P-7A program washed out than a disaster of larger dimensions emerged to reinforce those two principles. As the centerpiece of its new combat aircraft development was a new, spooky-looking, triangle-shaped attack airplane, the A-12. Designed of mostly black composite materials for a stealthy low-radar signature, like the better-known and even more controversial Air Force B-2 flying wing bomber, the A-12 had grown pricier and pricier. Two contractors, McDonnell Douglas and General Dynamics, were teamed in the program, and the rumor in defense circles was that neither was willing to talk freely to the other about the technology.

Whether government program managers in the Pentagon were in better touch with their bosses than the contractors became an embarrassing question. True, the Pentagon had returned badly needed responsibility and authority to the working program manager. Unfortunately, contradictory forces were at work at the same time. Until the awful truth emerged, the A-12 had been portrayed as a program on target. In fact, the A-12 was overweight, behind schedule, and at least a billion dollars above its cost target. This time the Navy tried to hang in, but the Secretary of Defense himself cancelled the program, leaving the government demanding money back from the contractors and the contractors suing the government to cover their losses. McDonnell Douglas, frequently at the top of the list of largest defense contractors, had to plead with the Pentagon for an advance on its other defense business because its cash position turned so dicey after the A-12 went down the tubes. So concerned did the Pentagon become about the impact of downsizing that it ordered a review of the financial condition of its top contractors.

Why had the Navy tried to obfuscate the financial bloating of the A-12? For the same reason, it turns out, that the Army grieved over lost programs at the Raleigh meeting. If a service confesses that a program is going badly, if it tries to temporize with program delays for fixes, a no-win dilemma ensues. Decisions escalate to the level of the Secretary of Defense staff, the controller particularly, who holds back funding under the guise of further program definition in which technical or financial feasibility of a weapon is studied all over again. Meanwhile, lots of meetings go on of the Defense Acquisition Board, the high-level service and defense secretary staff group responsible for Pentagon buying decisions. These meetings are run by the defense acquisition czar, who may say: "Oh, by the way, this program or that program has too much risk. Build more prototypes to prove out the concept or do more testing, take a few more years of development." Doing so raises costs and the money comes out of the service's weapons budgets. Production quantities are down by this time and the price per copy is up.

Then the analysts move in. They maintain a shop in the defense secretary's labyrinth called program analysis and review. This is a remnant of the Pentagon "whiz kids" introduced by Robert McNamara into the Pentagon in the 1960s to look over the shoulders of the services in their buying. While the function fell into disfavor under later secretaries, it regained stature when Dick Cheney moved over from Congress to become secretary of defense. His distrust of the services, particularly the Air Force, was evident, and program analysis, independent of the services and an available watchdog, became a handy bulwark. So in come the analysts who can argue: "You people are not serious about this program. You've done all the research and development and you're only buying X number of systems. And gee, look at the unit costs. It's gotten so bad I don't think we should even certify to Congress that the weapon remains cost-effective." Hence the no-win dilemma: confession of trouble means the service either will have to kill the program itself now or let the defense staff do it later.

When program analysts argue that the program is not serious, they add that, instead of having ten unhealthy programs, it is better to have five healthy ones. That would be well and good with the services—except that after termination of an unhealthy program, the DOD controller scoops up the money. This can't even be applied to a surviving healthy program. Not only does one service face losing these dollars for itself, but worse: they also could go into the pockets of a Pentagon rival.

Out in the wild blue yonder, the Air Force was having its own troubles. As peace broke out, the House Armed Services Committee tried, in the spring of 1991, to zero out the funding for the USAF's top-priority, $850-million-per-copy B-2. While the money came back in the Senate, the prospects for all 132 airplanes of the original program faded with every dropping brick in the USSR. To John F. Krings, who spent a couple of years looking over the shoulders of service program managers as the independent test director whom Congress forced on the Pentagon in the Reagan era, the political gamesmanship was evident. Representative Les Aspin, the Wisconsin Democrat who runs the House committee, in Krings's view, began to waffle after the budget markup on Capitol Hill. "From no B-2s," Krings said, "Aspin seemed to be moving toward, well, how many? And should we make the decision this year? Maybe we ought to keep funding it or maybe we ought to wait another year. I think what Aspin was trying to do was show he could get the entire United States Congress, or at least the House, on his side versus the president. Aspin owns the house and he will give back any money he wants to."

Aspin did another politically shrewd thing with the B-2, Krings said, "by putting it into the bank and passing all the money out. He took $2.4 billion out of B-2 and gave $100 million to this program or that one." Rather than terminate production of the Air Force's top-line fighter, the F-16, as the administration wanted to do, Aspin extended the F-16 another year. In the same way, the Air Force F-15 got another year. Then $800 million went to the radical tilt-wing V-22 troop carrier the administration wants to kill. Money went to the Navy's F-14 fighter that's been in production for two decades, another administration termination target. The Army's M1 Abrams tank got a six-month reprieve. "So everybody plans on another year of this and another year of that and another year of this," Krings said. "Fort Worth (where the F-16 is built) is happy and St. Louis (where the F-15 is built) is happy, and everybody's happy. If the Senate wants to do differently, it will be taking that program away from someone, even though that someone never owned it. Everybody will hate the bank, not the B-2 as a weapon, but the B-2 as a bank."

That left the B-2 on the edge of limbo, with the Air Force trying to sell its usefulness as a conventional bomber. By then it was obvious the final number was not 132, or 75, or even 50. Aspin himself said as much in a speech shortly afterward: that the issue is not whether to have a B-2, but how many. Maneuvering aside, then came the failed Soviet coup against Gorbachev, disappointing stealth testing of the first B-2s, and the idea to fold USAF's strategic nuclear command into a more generic structure. Congress was ready to trim the program to 15 airplanes at the end of 1991. Finally, the administration itself capitulated. The Bush fiscal 1993 budget proposed to stop the program at 20, by coincidence possibly equivalent to the inventory accumulated at that time that the government would have to pay for anyway. Five more B-2s do make a difference to the Air Force, for it will have enough spare aircraft for maintenance layups and training to keep two full squadrons operating. Even after Bush gave ground, Congress continued to try to trim the program further.

The advantage of the B-2, despite its price tag, is that it is cheaper per target. "Leave the bombs in California where they were built," Krings put it. "Don't bring them to Saudi Arabia. With the B-2, how much airlift do you need? Use the B-2 as a tactical, conventional attack airplane." In that concept, an airplane conceived as a specialized Soviet mobile missile killer becomes an all-purpose attack airplane. As defense changes, past distinctions for different kinds of airplanes, fighters, bombers, and interceptors will change, too, into platforms for delivery of various kinds of smart bombs or missiles or collection of intelligence. Hence the consolidation by the Air Force into new kinds of composite wings.

This is the way other expensive Pentagon weapons programs will play out as Congress continues to try to turn the military budget into a bank. Entailed in the B-2, in the Navy's canceled airplanes, in the Army's fretting in North Carolina over vanishing procurement dollars is an epitaph to the way the country's defense has been run for 40 years. Those years saw massive investment to build a huge industrial complex to support the waging of the Cold War. Now the war is over. Policy makers, politicians, and the military must grapple with how much expensive weaponry is needed for the future and how massive a support complex is needed to build them.

Both the United States and the Soviets strained themselves to carry the Cold War burden before the East Bloc rupture came. Unfortunately, the last ten years of the cold war coincided with a borrowing blowout for the United States and, in retrospect, a step toward bankruptcy for the Soviet Union. Now it's the morning after.

Is It Really Over?

How bad is the hangover? Between tales from visitors to the Soviet Union and the nightly news on television, there is a portrait of a dismembered empire strapped, nearly broke. For the remnants, the Cold War is lost and a new economic battle to wage is next. In hock as it is, America is having a hard time working out a Soviet Marshall Plan for the 1990s, aware as it is that such may be essential to keep Eastern Europe in the free market camp.

More comprehensive arms control treaties are on the way, anticipated by George Bush's nuclear standdown in 1991. But the Soviet Union's nuclear arsenal still existed as the United States headed into defense downsizing, with the new commonwealth states asserting rights of ownership and control over those in their territory, a force potentially capable of a devastating nuclear strike against the United States. For its part, the United States has not done away with either its submarine ballistic missile arsenal or its newer, long-range, land-based missiles. Nevertheless, progress in the standdown is clear in termination of a new U.S. small ballistic missile and cutbacks in the Air Force Advanced Cruise Missile. Cancellation is probable of the bulk of the Navy's new Seawolf attack submarines designed to prey on enemy ballistic-missile-launching undersea boats, though, typically, there were efforts in Congress to put money back into the budget. Certainly, the United States will have to maintain some kind of nuclear force. Just how much is hard to gauge when Soviet survivors are too disorganized to mount any kind of military campaign with the weapons they have even if the hard-liners came back to power. Although the dismantled Soviet threat is fading, some of the new states have begun to waffle on promises to destroy nuclear weapons, a signal that the threat has not vanished.

If the Soviets are in disarray, economically and militarily, what about the Saddam Husseins of today or tomorrow? There the experience of the Gulf War cuts two ways. The U.S. government is talking about smaller, more lethal, rapid deployment forces as its military future—as it should be. That is a nod to Saddam Hussein, to the concept that surprise shocks in the Third World are the kinds of military threat this country will face in the twilight of the twentieth century and the birth of the next. Valid as that notion may be, the Pentagon is having to justify once again spending billions of dollars on a new advanced Air Force tactical fighter or a Navy attack aircraft in place of the cancelled A-12 on the grounds that the United States has to field weapons better than those it has now, which, in turn, are better than anything the rest of the world can field. Skeptics wonder how much better weapons are going to get in the rest of the world. Well they might, considering the lower price for modification and updating of existing equipment that usually is rugged enough to last for decades at low peacetime usage.

Visions of future regional war are coming out of Pentagon-supported think tanks. Much as the military-industrial complex delighted in the quick victory wrought by high-tech weapons in the desert, there was relief, too. One lesson is already clear, Saddam Hussein made a decision that lost him the war long before allied forces fired the first Tomahawk cruise missiles into Baghdad from warships in the Persian Gulf or launched the stealthy F-117 strike fighters to take out Iraqi command and control centers. American forces in Saudi Arabia were at the end of a 7,000-mile-long, initially tenuous supply line and vulnerable in the first three or four months of deployment. If Saddam Hussein had believed the allies would eventually strike, he could have attacked at that touch time—and the precombat visions of heavy American losses would have been correct. Six months to get ready for a fight won't be allowed the next time.

Contrary to conventional wisdom before the war that air power could not win by itself, air power almost did. That, too, will affect future U.S. force planning and the kinds of new weapons that filter

through the budget strainer. Another precursor of the future for
U.S. defense is the way the war was managed. George Bush de-
serves credit for that. Reversing a trend of two decades in Washington, he gave the military forces a clear task, then let the commanders fight without micromanagement from the White House or
the Pentagon. It marked a return to the way wars once were fought
until the new technology of instant worldwide communications
gave headquarters the opportunity to direct tactics from a Pentagon
war room. For that matter, the political rationale of the Gulf War
was a rerun of the lessons of World War II: If we had stood up to
Hitler at Munich, history tells us, there would have been no World
War II. Therefore, stand up to Saddam Hussein, though the Iraqi
leader turned out to be no Hitler of the 1990s.

Retrospectively, the Gulf War looms as a precursor of the near-
term military threat to the United States. Early in 1992 the
Pentagon put out for comment new guidelines for developing
grand strategy that recognized that threat. More political than
military doctrine, they espoused a friendly domination by the
United States as the single surviving global power, friendly as
long as no other rival bestirred itself. Implicitly they recognized
a post–Cold War era of less emphasis on outright force and,
under American aegis, more competition using diplomacy, economic and trade policy, and technology and industry as weapons.
But this vision of the United States as the surviving superpower
of the Cold War also postulates that the United States will not
brook any overt challenge from its former enemies turned allies,
Japan and Germany, and will not stand by watching the rise of
a new nuclear power of any stripe. When a common foe disap-
pears, the possibility is accepted that allies may quarrel among
themselves. The implication: if the United States loses the technoindustrial contest, force might be an alternative.

Together with the traditional war game, what-if scenarios military planners undertook at the same time, the process was interpreted widely as justification for Base Force Concept levels of about
a million and a half uniformed people rather than a new vision of
national strategy. Perhaps for this reason the guidelines disappeared
into the clouds in policy councils. Yet they represented an A-for-
effort stab at trying to define the future military threat at a time
when that threat was vague, to be determined later.

Coming: Another Bow Wave?

Given that defense will shrink—and the political pork barrel process makes for a reservation—and given that the competition for fewer Pentagon dollars will leave some contractors foundering, what happens to the infrastructure built during the Cold War to support the military?

Defense cutbacks habitually happen first; then comes the search for plans to revive defense manufacturing if the old or a new Saddam Hussein emerges. This was certainly the case in the early 1970s. Like the federal response to layoffs, planning for preservation of a defense industrial base was ad hoc. By the end of the decade, Congress was holding hearings on the dissolute state of the defense mobilization base, underpinning the Reagan defense buildup of the next decade.

More than industrial base was involved. To meet budget ceilings, the military was deferring maintenance, modification, and modernization of equipment it could not afford to replace. Either the Pentagon would be faced one day with a too-long-deferred and unaffordable bill for repair and refurbishment—the bow wave—or combat readiness would deteriorate. Both came to roost in the Reagan administration, which in its first budget doubled the defense increase inherited in the last Carter administration budget to raise the combat readiness of the U.S. military.

Vestiges of this same dilemma are resurfacing now. To some extent, though, the problem is self-created. On one hand, the services are told by Congress they still need do a better job of using the equipment they have before starting on new research and development programs, that before inventing fabulous new things, they ought to see to what's already there. Or the Grace Commission excoriates the Navy for neglecting its rusting, mothballed cargo fleet at a time when the nation is short of sealift. On the other hand, the military, Soviet as well as U.S., tends to be a pack rat. Now retired Army General Maxwell Thurman is campaigning for ruthless retirement and replacement, butting heads with old habits of keeping junk in the inventory and then complaining about the high cost of keeping it working. And the Air Force, projecting it will be down 2,200 aircraft from its mid-1980s peak, got rid of 1,000 in just the first two years of the 1990s.

One battle in the Army came over its old two and one-half ton truck. It would cost $4,900 a truck to bring them back from the Middle East, a truck soldiers don't like and which, they gripe, doesn't perform well. So why bring them back? Why not dump them there? Decisions like these are matters of priority and may depend on the personal priorities of the political transients in top Pentagon posts rather than an overall service policy. The Joint Chiefs of Staff in the 1970s elected to put money into new weapons development and risk the bow wave in deferred maintenance. In the 1990s the secretary of the Army killed a program for a new recovery vehicle that some in Congress thought the service needed, a need he recognized. But on a visit to Fort Carson, the Army's winter training center in Colorado, he saw barracks built in 1960, with no hot water downstairs and no lights. He had to decide whether funding refurbishment there was more important than buying a new recovery vehicle. It was a case of using dollars for operation and maintenance rather than procurement of new equipment. The services argue that the latest savings they were able to squeeze out in force reductions were taken away from them by the upper echelons in the Defense Department to make points with Congress. If new equipment buying is deferred, if old equipment stays in the inventory without refurbishment, there will be another tidal wave coming in deferred modernization costs.

While the Desert War was still fresh, George T. Singley III, deputy assistant secretary of the Army for research and technology, expounded on an analogous problem: the defense industrial base. "The defense industry is going to shrink," he said then and the Desert War did not change things. "It's just a matter of how much and how fast."

When defense manufacturing slims down, it will affect the broader reaches of U.S. military-industrial supremacy embodied in independent research and development (IR&D). Independent research and development, is a cost-sharing deal between the Pentagon and its contractors designed to encourage innovation in military equipment technology—though shrinking defense companies would like to see the government carry the whole load in the future. It is analogous to the percentage of sales commercial companies allocate as overhead to research. "Since independent research is percentage of contracts," Singley said, "and since there are going to be fewer contracts, inevitably there will be less money for IR&D. We've already seen how so much of independent research is becoming almost B&P like." Translated, that means what is supposed to be research money is plowed into bid and proposal work to win new contracts, not to innovate. As Singley added: "Independent research projects are beginning to look more and more near-term and less and less far-term, the contractor thinking, 'What can I do to enhance my position to win the next contract.'" Once the Pentagon could let independent research cover technology it liked but didn't want to pay for, in effect capturing industry's work. In the future, independent research may not be that robust.

A robust defense industrial base has a practical side. Foreign sales are possible and these, in turn, stoke the fires to keep the domestic defense base warm. With a defense industrial base in place, design teams are kept together. Otherwise, with overall reductions in procurement and engineering development, Singley went on, design teams get disbursed. Ideas about survival with research alone to the contrary, Singley notes that the infrastructure of the defense industry, whether it's rotorcraft or ground combat vehicles or electronics, is really maintained by development and production lines. "If procurement monies go away," Singley said, "how are you going to maintain your engineering capability? Just having a bunch of scientists and not having anybody who understands production engineering or manufacturing processes or how to translate research into an item that gets into the field. . . ." So there lies a quandary.

As the reality of a smaller defense industry solidified, ideas on how to remodel weapons acquisition proliferated around Washington. Virtual swords was one idea, using advanced modeling and simulation to substitute for developing or buying actual weapons. Another idea in various forms is to keep research teams alive and active, with an option for production or building prototypes: in effect, preserving engineering teams and forcing companies to learn the difficult art of how to make money on research alone to survive. A variation is to carry weapons through research and development but keep the technology on the shelf until it might be needed in national mobilization.

Exactly that option was selected early in 1992 by the Pentagon, putting money into less expensive research and development and then quitting with a few prototypes. That means no big-dollar production. While prototyping, making a business out of research and development, keeps design teams alive and helps budgeting, it also means further erosion of the defense—and U.S.—manufacturing base. If a mobilization base is defined as a manufacturing base, then research and development prototypes don't keep that base warm.

Yet there is shrewdness in the Pentagon in going the prototyping route. This might even be the answer to a program manager's prayer. Weapons programs now are enmeshed in a complex system of regulations and legislation that slow down the development cycle and, in industry's view, drive up costs as much as 30 percent. Technology demonstration prototypes not intended for production can circumvent this cumbersome apparatus. Like the Air Force F-16 fighter of the 1970s, which started as a low-cost technology demonstration, prototypes have a way of slipping into full-scale production, but in a way good for the country as well as the Pentagon. They are cheaper and proven.

Research and development directly funded by the government as projected in budget planning does reflect the prototyping decision. It does so by holding level: $37 billion actually spent in 1989, $36.4 billion estimated in 1997. These estimates do not allow for inflation. Hence $36 billion in 1997 is actually a cut by whatever inflation factor is assumed, but 10 percent ought to be a minimum. Furthermore, long-term defense planning figures traditionally overestimate what Congress eventually appropriates. There is more to this story. Pentagon research and development has actually escalated from a $33-billion low in 1987 to a forecast peak at $38 billion in 1994 before tailing off again. Big programs like new Air Force and Navy fighters are behind those numbers, and Congress is not likely to allow bigger research and development funding without wondering exactly what is going on.

How Much for Defense, Anyway?

As low as the visibility is for forging future U.S. military posture, the jousting over budget size is understandable. Assuming that the threat to this country has fallen at least by half, a 50 percent reduction in defense sounds reasonable. Things are not quite so simple. First, the military needs a certain core size in terms of people, bases, and equipment to be effective and credible. That size might be 60 percent or 40 percent—or some other number. How fast size is trimmed is also critical. Five years is short term, in the kinds of lead times the military needs to train people and equip them. Timing also depends on how much pain is tolerable, in employment dislocation and technology and disruption. Affordability is another factor, in light of the voracious federal deficit. Lacking a clear and present danger, it's like insurance: the nation can afford as much defense as it wants—or as little.

Despite the Reagan defense buildup, defense in 1990 was 5.5 percent of gross domestic product, a lower share than it was in the peak days of the Cold War, when it ran at 9 percent or 10 percent, in the late 1950s and early 1960s. Budget projections in 1993 called for a drop at the end of the five-year planning cycle in 1997 to 3.6 percent of domestic product. Defending military budgets on how tiny a share of the national output they take may be acceptable in times of crises, but not when perceptions change. Pentagon planners recognize that point in the fluid nature of the Base Force Concept for future military sizing. How can the Pentagon expect to command 1990's one dollar out of every four in the federal budget to meet a threat to be defined later? Or, for that matter, one dollar out of every six projected in 1997? There is enough public doubt that World War III is anywhere near imminent to make that a tough sell.

Prudence nevertheless suggests caution in cutting until the dust clears in the Eastern bloc. Euphoria at the end of the Cold War and the much diminished possibility of another global war do overlook a new kind of regional threat, albeit ill-defined. Retired Air Force General James Abrahamson, who directed Ronald Reagan's Strategic Defense Initiative, believes there will be the threat of small-time regional wars exploding into big-time nuclear or chemical warfare as more nations develop advanced technology for weapons. Terrorist groups may lay hands on the most lethal of modern weapons technology. Hence the U.S. near-panic over Iraq's undestroyed nuclear facilities, considering that perhaps fourteen nations have now acquired or are working on ballistic missiles, on means of delivering chemical or, someday, nuclear weapons.

Difficult-to-detect explosives such as the one that destroyed a Pan American 747 over Scotland are lethal enough. Chemical or nuclear weapons and the means to deliver them, in the hands of unstable, shadowy organizations, could become a global nightmare. "From this," General Abrahamson adds, "will come demand for improved tracking of the tools of warfare. Not only will this include monitoring of the reduction and destruction of weapons, but it will also entail insurance against the ability for rapid marshaling of those weapons that remain and could be deployed against reduced defenses. Tracking and real-time data on location of forces will be a key to future stability."

Future threats to increasingly open societies will materialize in their dependence on electronic communications networks and computing—modern society's technical infrastructure hostage to a kind of technical terrorism of splinter groups by their own, not national, agendas. General Abrahamson points out that a communications infrastructure might become a new kind of tactical target, not a railroad bridge or an airfield. Biological war of another kind, the launching of software viruses at increasingly complex global computing networks, could lead to a similar kind of strategy. Relatively small-scale computer viruses have already made very clear how disruptive this threat could become. Nuclear standdown by implication visualizes the terrorist, splinter group threat. All this will put priority in time of defense budget shrinkage on more adroit national intelligence gathering.

Fluidity means the Base Force Concept is designed for the battles ahead with Congress over size of the military. That give-and-take will bring out the best, and the worst, in the American political system; the best in that all the tradeoffs will get a fair hearing; the worst in that pork barrel or bureaucratic imperatives might outweigh cost control. Defense could become too much local pump priming. In the end, the military restructuring will be a mixed picture in the total budget sense, as inflation and people protection lead to disappointing reductions. A peace dividend will come in less budget growth, in less spending than would otherwise be the case, of funds difficult to spend elsewhere. As has happened already, weapons buying will be at the end of a cracking whip, disproportionately trimmed. Thus the military adjustment will be easier than the counterpart restructuring of defense industry, which has to decide whether to stay in defense or try something new.

3

Surviving as a Defense Company

Defense contractors were under no illusion that peace might break out one day, though the day-to-day scramble for new contracts over-whelmed planning for their inverse Armageddon. Not long after Richard A. Linder in the mid-1980s took over the Westing-house military group in Baltimore, which builds high-technology ground and airborne radar, he dropped Defense from its name to call it just the Electronic Systems Group. Foreseeing the seeds of reaction in the Reagan defense buildup, the Group had aimed by 1995 to leverage its defense technology base to at least 50 percent of its business outside of the U.S. Defense Department. By 1986, the fraction had reached 19 percent, and 1991 sales were better than halfway there, at 30 percent. How much real diversification is embodied is open to argument, such as adapting the Group's au-tomatic target recognition technology to automatic character recog-nition for the post office or military radars or infrared sensors for Coast Guard and customs drug interdiction. Civil, yes, but still drinking from the government well. Furthermore, Westinghouse is not the only defense company wooing the post office, which is chronically interested in automation and chronically short of money. Another defense contractor, Martin Marietta, won a $19 million postal contract for bar code readers.

About half its nondefense business is with other government agencies, but Linder's Group is looking to apply core competencies to the likes of home security or smart highways. And it has equipped over 100 airports in the United States—and countries like Poland—with a dual-use version of its military ground-based radar for air traffic control. Sticking to core technical expertise is the trick to transferring technology out of defense; otherwise it becomes a risky excursion into unfamiliar territory, Linder says.

When defense contracting dropped in the early 1970s, military contractors made the same tracks, not for commercial markets but for some place else in government: the National Space Agency, the Environmental Protection Agency, even the Housing and Urban Development Department. Stanley Barton, a British-born chemical engineer who ran ITT's Rayonier Division before retiring a few years ago, once had the job of converting that conglomerate's defense-commercial mix to 50-50. Defense divisions knew how to market to the government, he recalled, so their response was to pursue agencies other than defense. Less dependence on government, period, was what ITT's masters wanted, and after 18 months Barton recognized it was a futile exercise. ITT did get its mix more into balance before he took on other chores, Barton said, but mainly because its defense business fell into line. But there may be a diversification of a sort in disarmament. Lockheed, whose aircraft and missile work put it in touch with nuclear weapons technology, has formed a joint venture with a nuclear fuel supplier and an ordnance builder to dismantle such hardware. The market may be in the former Soviet republics that lack the technical talent for a delicate job.

The Last Time Around

To get an idea of what might happen to people and companies when defense shrinks severely, take a retrospective drive through Cocoa Beach some time in the early 1970s, the little town in Florida that originally was the bedroom for the Air Force ballistic missile test range. Every other house seemed to have a For Sale sign in its front yard. As the U.S. space program burgeoned, Cocoa Beach oozed along a sand barrier beach into a collection of motels, restaurants, and the Florida coquina limestone stucco one-floor houses where the engineers lived. When there was a space launch, it swung. Everyone from high-domed scientists to journalists from all over the globe crawled through the bars, sat through briefings, sunned on the beach, or waited hours on hot bleacher seats for the next giant firecracker to light off. When the launch was over, the town turned back into a sleepy little southern pumpkin.

Trouble was in the early 1970s Cocoa Beach was headed for more than a short nap. Frenzy was tailing off in development of ballistic missiles. The third-generation Minuteman intercontinental nuclear missile was in holes all over the Dakotas, Wyoming, and Montana, and nothing else was coming along. Defense budgets were squeezed for a peace dividend as the Vietnam conflict faded. Worse for Cocoa Beach, the Apollo manned lunar landing program that built the Kennedy spaceport was waning. President Richard Nixon had decided to stop gathering moon rocks and dust. The last mission, Apollo 17, was the only one launched at night from the Kennedy center to the north of Cocoa Beach—an apt end, the Saturn booster torching the night sky with a greenish-yellow brilliance that looked like the second coming itself. Back in Cocoa Beach, though, Chamber of Commerce meetings convened to think of ways to attract other kinds of business to the town, those not dependent on government largess and changes of mind.

Times were even tougher on the West Coast. Well-paid engineers in defense companies were pounding the bricks, laid off as military procurement dollars dwindled, as they are fated to in the early 1990s. Commercial manufacturers told them to get lost, that their defense cost-is-no-object mode had no place in the real world of markets and competition. Seattle was a special case. Not only had vanished military dollars hurt Boeing, the region's largest employer, but so had the end of Apollo. Boeing built the first stage, a cluster of tankage and rocket motors, of that monster of a launch vehicle. On top of that, Vietnam War–generated traffic ceased for the airlines. Boeing went for over a year without a sale to a U.S. airline, and by the end of 1971 it had laid off 53,000 of the 83,000 employees it had in 1968. Two of them paid for a billboard near the airport with a picture of a lightbulb hanging from a wire and the lettering: WOULD THE LAST PERSON TO LEAVE SEATTLE PLEASE TURN OUT THE LIGHTS.

"A lot of engineers in Seattle wound up driving taxis and pumping gas," recalls Ray Waldmann. Now director of government affairs for Boeing's Commercial Airplane Group, Raymond J. Waldmann saw the disruption from the other side. Then he was part of the Nixon White House domestic council, working for John Erlichman, who went to jail after the Watergate disclosure a couple of years later. "Erlichman, of course, was a local boy from Seattle,"

Waldmann explained. So in 1970 and 1971 the administration decided to take on a program to retrain laid-off aerospace engineers because it was a West Coast problem, but, Waldmann said, "it never had the impact or the expected results."

Labor Department seminars told engineers how to get jobs that didn't exist, how to write résumés, how to interview. There was an environmental program. Obviously environmental, waste management was a coming thing. Many inspectors would be needed for the Environmental Protection Agency, or so ran the common wisdom, Waldmann recalled, but, he added: "Any government agency, anything anyone could dream up" might become a program.

"Primarily a West Coast problem" was a delusion. Defense or aerospace concentrated in areas like Seattle and Los Angeles or Boston were in a depression while the rest of the country struggled along with a mere recession. Everybody was diversifying. Computers were popular. McDonnell Douglas, another big defense and aerospace manufacturer that started an operation called MacAuto, bought a service bureau operation, Tymeshare, and put a lot of effort into the business for 20 years. Environment, garbage disposal, desalinization, anything with a new technology spin was considered. Some defense contractors took a shot at factory-built modular housing. Lots of effort went into systems analysis of urban problems. Engineers produced schematic diagrams of how urban functions interrelated, something city planners knew already.

Vertol, a Boeing subsidiary in the then fast-disappearing business of producing military helicopters, won a contract to build trolley cars for commuter lines in its hometown Philadelphia. Again, this was a business dependent on what turned out to be the tenuous willingness of the federal government to advance money to states and cities that couldn't or wouldn't pay for mass transit on their own. Bridgeport, Connecticut was pinched as the assembly line at one of its major employers, another military helicopter builder, Sikorsky, emptied out. Sikorsky also tried to diversify into the rail business, building a technically innovative turbine-engine-powered, high-speed train. The New Haven Railroad, itself broke, ran one for testing and so did the Canadian national rail system. But the aerospace-engineered turbine train didn't survive. Companies found that they burned up a lot of good people trying vainly to make these kinds of businesses succeed.

As Ted Bluestein remembers it, a lot of the diversification and retraining was nonsense. Now retired from Hughes Aircraft, Ted Bluestein can look at those days with a sense of humor. Then with Raytheon, coping with cutbacks in Boston, Bluestein remembers with a hollow laugh, "You know, I'm from the government, I'm here to help you. Well, in Massachusetts, we got lots of government help. It was awful. They were retraining people left and right. But all sorts of nonsense stuff." Rapid transit, pollution control, and even education—teach them to be teachers.

A logical question now, if the country is headed back into this same dilemma, is whether any of the retrained engineers ever got work in their new fields? "Not that I know of," Bluestein said. One friend of his at Raytheon did what scores of his colleagues did, waited it out, maybe doing a bit of consulting, until the defense industry recovered in the late 1970s. "One guy started a hobby shop in Concord," Bluestein recalled, but he didn't know whether it survived. "Let's face it. Even a good small retail business isn't going to give you an income like an engineer will get from aerospace industry." Another man, a sports car buff, opened a garage in Bedford that lasted a couple of years. "There were a lot of guys like that," he said, "but I don't know what happened to them."

Sadly for the laid-off engineers, and those without formal degrees were the first to go, nobody seemed to want them either in industry outside of defense. Commercial manufacturers were hiring engineers, at least in small numbers. They didn't want engineers fresh out of the defense industry. Therein lies a truth about the defense industry and a foretaste of the difficulties that lie ahead. Defense engineers, at least in the popular mind, worked with limitless money. Bred in a cost-plus environment, designing for the ultimate requirement, combat, they were indeed misfits in the commercial world that survives on trimming expense and shaving margins and could care less about military specifications.

Defense companies fared as badly trying to change corporate careers. LTV, the renamed, merged corporation that grew out of the old Vought fighter aircraft builder, was helped into the automated ground transport field when its hometown built the new Dallas-Fort Worth airport. Although there were recriminations over performance and contract terms, LTV's automatic railcars still shuttle between terminals at the Texas airport. Robert Parker, a Pentagon research manager and later Vought's research and development chief, put a final sentence to the tale: "We lost $35 million on the program." Rohr, another defense and aerospace contractor, dove into rapid transit, building cars for the Bay Area Rapid Transit system in San Francisco that were rife with the troubles that any brand-new technology has when first introduced into service. Boeing, with $120 million in federal money and the support of the late Representative Harley O. Staggers, Sr., built a personal rapid transit system for the University of West Virginia campus at Morgantown, one that also was burdened with these kinds of introductory troubles. Computer control turned out to be high-tech overkill at the time when a simpler elevator-control system would have been more reliable and cheaper, but the system still runs today.

All these attempts at conversion out of the defense business have similar implications for what the 1990s hold for military contractors. Their high-technology civil equipment worked after the bugs were fixed, but contractors, used to having costs covered by the Pentagon even if, as usual, they overran estimates, did not have the capital, the staying power, or the inclination for development of civil markets. They tried to introduce new technology, but it was technology like Sikorsky's turbine train or Boeing's commuter system that the market didn't need or want to pay for. Defense contractors could convert technology, but they did not have the commercial experience or knowhow to make money with it.

Federal policy makers were caught short in the Vietnam downturn. Trying to maintain a fiscally restraining federal budget surplus when it took office in 1969, the Nixon administration cut $4 billion out of what seems today a miniscule $192.9 billion total. Of that, $1.1 billion was from defense. When uncontrollable items like interest on the debt overran estimates, the administration cut defense $3.5 billion more.[1] Expecting a full-employment budget surplus in 1970, the administration instead saw tax receipts fall $5 billion from the year before. Defense weapons buying dropped steadily from a Vietnam peak of $23.9 billion in 1969 to $15.2 billion in 1974 although overall defense budget reductions were only half that. Other spending spurted, including $1 billion more for unemployment compensation, exacerbated by 1,100,000 million defense layoffs.[2] Driven by defense cuts, unemployment rose from 2,832,000 in 1969 to 5,016,000 in 1971.

In real terms, defense weapons procurement began to drop again in 1987, falling 12 percent from $88.5 billion to $77.9 billion in 1989, and then was kicked upward by the Gulf War to $80.5 billion in 1990.[3] From a 1991 Gulf War peak of $82 billion, procurement was forecast to sink 18 percent in 1993, to $67 billion, in current dollars. Since 1990, the White House Council of Economic Advisers said in its 1992 report, a much larger downsizing has begun to affect defense industry employment, falling, according to Commerce Department data, from 1,253,000 million at the end of 1989 to 1,105,000 at the end of 1991—with more to come. If that two-year trend continues, defense employment will be down to 853,000 at the end of 1995 and 655,000 at the end of the decade, half what it was at the start. "The United States has accommodated reductions in defense spending before," the 1992 economic report said. "But the transition is never easy and, in fact, is costly in the short run. . . . ." Not only is this costly in retraining and retooling, but it added: "Local economies where defense industries are a primary source of employment can experience significant disruption." Clear enough that the experience of the 1970s has not been forgotten.

As for laid-off defense workers, Boeing's Ray Waldmann thinks the economy of the 1990s can absorb people, especially trained military people, more easily than the economy of the 1970s. The bad news is that there will be a larger government increment to the layoffs now. Transition from government to private industry is harder than transition from defense industry to commercial. Not that there aren't opportunities in reducing the defense claim on the economy. "That was the argument in the early 1970s, too," Waldmann recalled from his White House days. "Lots of people were saying it's about time we freed these guys. Let them start working on Tupperware or rapid transit."

In fact, they may go into something else that wasn't around in the 1970s. As Waldmann sees it: "All of these freed engineers and scientists are going to end up working for Sony or Toyota or Honda." Japan is hiring American scientists and engineers now and may see surplus defense staffs as more than just people who are competent to assemble things, but as industrial or design engineers or for populating research laboratories, tapping more American technology in the process. "Maybe," Waldmann said, "Japan will absorb them faster or better than our domestic companies will." In the meantime government retraining and relocation programs for defense workers are starting again at the Labor Department.

Coping with the New Reality

Layoffs and contract cutbacks of the 1990s will be larger than those in the 1970s, perhaps the largest since the instant deflation of the defense business at the end of World War II. An even bigger difference this time is the prospect for defense industry revival. Blockade in Berlin, war in Korea, and the open-ended Cold War with the Soviets always reenergized ebb-and-flow defense cycles through four decades. But with the remnants of the Soviets in disarray and Saddam Hussein beaten, a new crisis is hard to visualize. Defense industry is headed for slimming down, consolidation, and change on a scale not seen since World War II.

Two strategies have emerged in the defense industry as shrinkage has become national policy with a flavor of permanency. One is the approach of the profit-minded former astronaut, William A. Anders, who runs General Dynamics, which is projecting dropping as many as 30,000 workers. Looking at a relatively large defense backlog, Anders recognizes defense will become a dicier business but proposes to become more profitable in a smaller market.

On the other side are those in defense who are saying that things are not going down so drastically, but if big cuts do come they won't happen to their program. So, in Capitol Hill jargon, "they send up the corporate grin guys to the Hill" to sell the idea for the National Guard to buy one more year of production of a fading weapon. People there keep their jobs for one more year. Jack Krings, a former test pilot, Pentagon independent test director, and now a defense consultant, contends that the earlier they get out with money to do something else, the better off they are. "If," Krings said, "a company is willing to take an objective view of how it stacks up against the opposition, the chief executive may say, 'I'm going to get out tomorrow, because I look around, and I see that if anybody's going to be alive, it isn't going to be me.'" Contractors are all equal only when defense is giving out a lot of money. When government funding shrinks, some won't finish in the money.

As newspapers day after day chronicled more layoffs at defense contractors, defense contractors contrarily reported decent profits. Investors got soft soap about the future of the defense business, and, in fact, Congressional cuts have been more talk than reality. Typically, though, defense contractor finances look the best at the worst of times: when business is running out and cash, accumulating from past sales, has no place to go.

Budgetary illusions play a part, too. Reductions are coming first in defense budget authority, one of three broad categories in how the federal government accounts for its spending. This is money Congress allows the military to commit to new contracts for payment months or years later. Contractors can truthfully say they are not hurting much now, but that is only because they are being paid out of another budget category: outlays, spending authorized years before. Budget authority becomes the outlays of the future, and outlays will surely follow budget authority down. Besides, weapons development is a decade-long exercise. Programs deferred or budget reductions today hit industry many tomorrows hence.

While the Gulf War seemed at first a reprieve for weapons spending, that wasn't the way Mal Currie felt about it. Malcolm E. Currie had seen too much of the defense world to believe so, from both industry and inside the Pentagon as director of defense research and engineering before he became chairman and chief executive officer of the Hughes Aircraft operations of General Motors. Before the shooting started, Currie said presciently: "However Desert Shield turns out, it is going to hurt the defense industry, not help it. If we ever had a chance for a slightly more gradual soft landing, it's become much more uncertain at the very least."

By this, he meant the prospect of continued defense cuts into the 1990s. Desert Shield and the Desert Storm shooting phase stayed the personnel axe briefly, but, as Currie pointed out, the money to pay for the reserves abruptly called up to active duty had to come from some place, and he added: "Right now I can tell you the Defense Department is just in turmoil trying to figure these things out. They're canceling and stretching out programs. Instead of helping defense contractors, Desert Storm is going to accelerate that whole process and make it much more chaotic." So it did, in retrospect.

Meanwhile, defense company spokesmen like Don Fuqua, a former Florida congressman and now president of the Aerospace Industries Association, urged a gradual, orderly transition to lower budgets—but not disputing the fact that reductions were inevitable as the Soviet threat cooled. The pace in Currie's view has not been all that slow. "It is still pretty drastic from an industrial point of view," he said, "but at least it was something you could begin to plan around. In industry, we have very little to plan around right now." Again, that was before Congress and George Bush began drawing lines in the sand for defense funding.

So a repetition of the disarray of the post–Vietnam War winddown is clearly a possibility. Few in the Defense Department, to Currie, seemed to understand the problem to begin with from an industrial point of view. "So," he said, "the easy way out is to say, 'What the hell, we're a free capitalist society, we believe in free trade, survival of the fittest, supply and demand, let's just let the chips fall where they may.'" Dislocations initially, but all will calm down in a few years. Left hanging in that premise are those companies, like Hughes, that had been most responsive to the government, in buying hundreds of millions of dollars worth of machinery and equipment, building modern factories, spending for research and development. Those changes were forced on defense contractors who knew full well the government could take away contracts as fast as it gave them. As Currie worried, "They're left with this tremendous financial overhang that has to be written off somehow." So the most responsive are stuck with big overhead and are less price competitive. Mal Currie's apprehension is clearly reasonable: that changes in the U.S. defense budget will leave high and dry those companies that invested in defense as a business.

The first wave of employee reductions at Hughes was carried out rather generously and with less pain than the usual, with fairness to those who stayed and to those who left. These separations, as personnel managers like to call them, were the golden handshake variety of early retirements for those who likely would be leaving shortly anyway. "I saw this coming," Currie said. "In the last half of 1989, we created an accelerated retirement plan, and we took the head count down by 9,000 people, which was about 12 or 13 percent of the company. And the results showed. We're more competitive now than we were then, but it cost us a pile of money to do that." Though Currie tried to assure everybody of stability, except for normal attrition, for two or three years, the aftermath of the Gulf War Desert Shield and the Soviet collapse brought concern that Hughes would have to cut down a little further. "In the future," Currie added, "it will be just a matter of economics. We're not going to be able to be as generous as we were in the past. We won't have the money for it." This was prophetic, for a year later Currie canceled raises in 1992 for senior management and put off salary review for the others until the middle of the year. Then, when Currie retired in early 1992 as chief executive officer, he was succeeded by C. Michael Armstrong. By mid-year, Armstrong had accelerated an existing plan to lay off another 9,000 employees—15 percent of the Hughes work force of 60,000—and close plants, mostly in southern California. Armstrong came from IBM, a manager with broad commercial industry experience, not the usual Hughes career defense operative. As such it was a message to the huge Hughes apparatus that business as usual was no more, and that Hughes, as an electronics house, if far better able to try to find commercial work than the metal benders who build fighter aircraft or submarines or tanks. Commercial work, though, won't need all the high-priced help of a big defense contractor.

Other defense contractors are in the contraction and restructuring mode as well. Boeing's helicopter division and Sikorsky won a potentially large scout helicopter contract, which may keep jobs in Philadelphia and Stratford, Connecticut. Where there were two winners in a complicated flyoff competition, there were also losers. McDonnell Douglas Helicopter Company in a Phoenix suburb faced layoffs and later went on the block. McDonnell's St. Louis fighter factory complex, on the other hand, got a new development contract for the Navy's F-18. But then, with more Navy money going for F-18s, the other big Navy aircraft house, Grumman, became a loser—though New York congressmen protested forthwith. So the Long Island company, much smaller than McDonnell Douglas or General Dynamics, is going through its own version of layoffs, with a plan to cut 1,900 employees from its 25,000 workforce. There is no certainty that those are the end. As a fitting postscript, both budget and design problems are raising costs of a new F-18 version, which foreshadow cutbacks in how many will be built.

LTV Corporation, a steel conglomerate in bankruptcy court, cut its corporate staff headquartered in Texas by a third, then started a bidding war among other domestic and foreign defense companies by putting its Vought aerospace operations on the block. Like Bill Anders at General Dynamics, others think defense contractors can slim down and make money in slower times. At first convinced that a defense reprieve was unlikely, General Dynamics cut back research as well as personnel and for a while acted like it might even look outside the defense market for business. Then it decided to concentrate on becoming one of the survivors in defense. So it sold off its largely commercial Cessna Aircraft company instead. This was a special case, though, for Cessna is in the product liability decimated market of corporate aircraft building. Sale to another defense conglomerate, Textron, was a way for General Dynamics to slide from under $200 million in potential crash lawsuits.[4] And Textron, conversely, was looking for ways to expand into related commercial business.

Patching up Old Equipment

Of course there is another option for defense contractors and the military: refurbish and modernize what's in inventory. Retrofit is the name for it in the trade; it's not the crutch it sounds, and even a richer American military has had sizable service life extension programs for years. As aerodynamic innovation came harder and electronics became more important, aircraft became far easier to update than replace. Modern airframes can last half a century and it is relatively easy to fit new electronic black boxes and even a newer, more powerful engine in an old shell. Aircraft now are electronic and missile platforms, the platform itself not so important.

Foreign orders are another option for contractors struggling for business and for the military in keeping a warm mobilization base. The downside for U.S. weapons builders is that overseas customers, more strapped than the United States and melting down smaller force structures, will increasingly go the retrofit route rather than become a market for U.S. exporters. Sometimes the services themselves torpedo overseas sales, as the Navy did by giving ships away to Greece or Taiwan. For that matter, the remnants of the Soviet military formed private companies to sell off ships as its military disintegrated further.

Nevertheless, foreign orders can be lifesavers. General Dynamics, predicting its head count would drop by 10,000 in 1992 as the F-16 production phased out, got an unexpected order from Korea that staved off closing down the line. In the meantime, the Air Force plans to continue buying modified versions of the existing F-16 as the most budget-conscious way of avoiding a shortage of fighters until a new multipurpose fighter can be developed, if it is.

From famine, General Dynamics went to a 1990s form of a defense feast. For General Dynamics was a member with Boeing of Lockheed's winning team for the Air Force's Advanced Tactical Fighter, heralded as a contract worth $13 billion initially and $75 billion eventually. As the military's firmest new aircraft program so far in the 1990s, possibly the only one in light of defense slimming down, the winners can hope for survival in military aircraft contracting. Yet the advanced fighter, USAF's top priority program that exists as the YF-22 in prototype form, has the same baggage to carry as the Air Force's fading strike fighter replacement. Public reaction to the original contract award drew a continual question: Why does the United States need to spend over $100 million per copy for a brand-new air superiority fighter to replace the F-15 when nowhere in the world is there anyone capable of building anything nearly as good? When the single flight test prototype of the YF-22 crashed in the spring of 1992, it had acquired 75 to 80 percent of the expected data, but the psychological effect in Congress reinforced its opponents. So the Advanced Tactical Fighter represents a dilemma all the services will face. Should the Air Force gamble on the best in a YF-22 that is not due for production until the next century, or stick with modifying its existing F-15 fighters that the YF-22 will succeed—as the Navy did with the F-18? If peace dividend and deficit pressures eventually block production of the YF-22 and the F-15 assembly line is shut down, that sums to zero.

On top of the cool reaction to the advanced fighter lies another uncertainty: the new defense policy of concentration on research and development of new weapons into the prototype stage and avoiding production commitments. The YF-22 could stay as a handful of prototypes. Congress, looking for an elusive peace dividend, will be heard from on that issue.

Changing Old Habits

Forty years of Cold War military weapons research and development has evolved an unwritten dictum: custom-design it. Yet one of the smartest weapons of the Gulf War was commercially developed. Not a weapon at all, it was a palm-sized navigation set that used satellite bearings so that tank crews knew exactly where they were in a featureless desert, and it was relatively cheap even though developed to military specifications.

With this kind of success, with the squeeze on future military dollars, comes the question: Why all the emphasis on custom design? Customizing is expensive and time consuming, especially the way the military does it. Literally thousands of specifications, so-called Milspecs, litter the pages of contracts for defense hardware. On top of that are military standards. Where specifications spell out performance demands—the what—standards spell out the way equipment is to be built—the how.[5] Both are invoked in the name of ensuring that military weapons are rugged enough to stand the rigors of the field: temperatures well above 100 degrees in the desert, temperatures well below zero in the arctic, sand, snow, rain, ice, and, not the least, rough handling by the user.

But, observes Tom Wooten, who heads the Research Triangle Institute in North Carolina, which does frequent work for the military: "I know the military environment is a very rugged environment. But I've got electronics on my little boat that I bounce around 40 to 50 miles off the coast of North Carolina, and believe me that's a rugged environment: salt, heat, vibration. My electronics are not Milspec, but they're still highly reliable." His upgraded system included a sophisticated depth finder and an automatic Loran navigation black box and plotter costing about $700. "When I was a navigator in the Navy, I had a rack of equipment that was about the size of that cabinet," Wooten said, pointing to a credenza in his office. "That cost the Navy at least $100,000. And it wouldn't do half of what my $700 box does."

There is a message here: how long can the military afford the luxury of custom design as money dries up for weapons? As the Congressional Office of Technology Assessment pointed out, defense not only should draw more on dual-use industries, but it will also have no choice but to do so in some instances[6] "because the technology is ahead of what the defense world is building. Increasingly, leading-edge technology is developed in the civilian sector and then finds its way into defense applications." To be sure, the Pentagon has raised its use of commercial items by an order of magnitude over the last decade, and its use of commercial specifications instead of its own has grown commensurately. But commercial items are less than 10 percent of total procurement.

Technology aside, cost differences are sobering. At the Army's soul-searching meeting in North Carolina, Jacques S. Gansler put a chart on the screen that caused a ripple around the room. Not that the message was new. The data, from Motorola in the middle of the 1980s, have been controversial since:

Commercial versus Milspec Semiconductors
(comparable part for comparable environment)

	Commercial	Milspec
Part cost		
Bipolar digital logic	$1.67	$15.78
Bipolar linear	0.42	11.40
Reliability (failure index) (high is worse)	0.06	1.9–4.6
Lead time for new part	1–12 mo.	17–51 mo.

"We always knew we paid more and we thought we paid more to get better stuff," Gansler said. "But commercial components are more than an order of magnitude better than Milspec." The reason why is clear enough. High commercial volumes drive continuous process and product improvement, and hence costs down and quality up. Yet one more disconcerting fact turned up in the data. Lead time to manufacture a new commercial part is a third of the military's.

Buy a car today, Gansler added, "and there's a chip hard-mounted on the engine block. The environmental specification for that chip is more severe than [the comparable] military specification, missing only radiation hardening." If environments are comparable, and commercial parts perform better and cost less, in tandem with the estimate that the Defense Department spends about

30 percent of its dollars on electronics and will probably spend up to 50 percent in the next generation of equipment, the question naturally arises: Isn't electronics worth considering as an area for dual use, at least at part level? Westinghouse's Dick Linder agrees. "We use a lot of commercial components now," he noted, "but unfortunately bought under a different drawing with a different specification. They may come off the same line, but have to undergo different testing. We have to remedy this, and take a perfectly good commercial component and plug it right into a military circuit."

Gansler, like Mal Currie, came out of the Pentagon as an assistant director of research and engineering before becoming senior vice president of the Analytic Sciences Corporation, a Washington technology consulting firm. He was one of the cochairmen of a think-tank study—along with Senator Jeff Bingaman of New Mexico—espousing dual-use technology for the military. One of its more arresting sections deal with the aggravations of companies that try to sell commercial hardware to the government, among them Hewlett-Packard. Cofounded by David Packard, later deputy secretary of defense and chairman of the White House Commission that looked into defense waste, fraud, and abuse scandals in the mid-1980s, Hewlett-Packard was an early bloomer in California's Silicon Valley electronic startup greenhouse. Among other things, it pioneered the versatile hand calculator and the laser printer. What it tried to sell to the government was an advanced signal simulator known by the acronym FASS for Frequency Agile Signal Simulator, a black box it had developed for commercial sale.

The report said:

> Its capability to simulate a wide range of complex signals and to switch easily from one signal to another stimulated considerable Defense department interest because of the potentials for electronic warfare and improved radar. The off-the-shelf commercial unit price is about $200,000 versus a less-agile, customized (DOD-unique) system with a price in excess of $1 million. Because the Hewlett-Packard system is new, however, it lacks an established commercial price—based on prior sales—and hence cannot be sold to the government without provision of cost and pricing data, which the company is unable and unwilling to provide. The net effect is that technological solutions tend to be based not on who has the best approach, but on who can (or is willing to) cope with the paperwork.[7]

Cost and pricing data are only one element in the constellation of nuisances met by commercial companies dealing with the Defense Department. In the case of converting a commercial 747 jet transport into Air Force One, the flagship of the government's VIP fleet, the airplane that flies the president of the United States around the world, the report noted that it necessitated 1,800 pages of such data, plus potentially 5,000 pages more of backup information. All this simply for modifications. The commercial airplane itself was exempt from certified cost or pricing data. Both the panel report and the Congressional Office of Technology industrial base study pointed out the same fact: that commercial companies often would rather not, or even refuse to, do business with the government.

Naturally, there is another side to dual-use technology. Military stuff often enough does have to be hardened against radiation from a nuclear blast, or against just garden variety shock. Logistics is a long and complex train, and even minor components have to stay on the track. They must fit the same way every time, the function must remain stable. They don't tolerate frequent commercial model changes, because the military operates antiques as well as high-tech hot rods. So there is more than the not-invented-here syndrome to resistance to commercial components. Yet the situation is plain enough. Future military budgets will force the issue, force it toward dual use, toward more commercial components in defense hardware. The Army has the message, for in its New Training Helicopter program in late 1991 it set out to buy a low-cost commercial helicopter that could function as a primary, instrument flight, and navigation trainer without compromise to military effectiveness.

Attractive as dual-use technology may be in lean times, there are other potential problems for commercial companies. Protection of intellectual property, a rotund work for technical secrets and patents, is one aspect can run afoul the military's drive to control data rights on programs it finances. Another is the danger of getting tangled up in the U.S. military export control web. While the need is recognized for loosening these controls if Eastern bloc industry is to get the technology to survive in free markets, barriers will not evaporate. William Schneider, Jr., who learned the foreign military sales ropes in the State Department and now works in the export financing field, says that the system is headed the other way at a time when a lighter hand is needed on controls. Instead of concentrating on one big Cold War rival, the export controllers are now up against a more diffuse set of countries. "They have maybe two dozen countries they're watching," he says, "not the usual pure terrorist Libyas and Iraqs but others who are living on the edge, trying to break into the system to get things that might be mundane under other circumstances."

Dual-use technologies have an impediment in the military's procurement system: bias against plant integration. Manufacturers, though, want to build compatible defense and commercial hardware in the same plant. Jacques Gansler testified before the Senate Armed Services committee in the spring of 1991: ". . . Today the Defense Department cannot easily reach beyond its captive defense industrial base. DOD's business as usual virtually forces a separation of the private sector into two discrete economies: defense and non-defense." Contractors know by now that regulations covering everything from defense cost-accounting rules—a far more difficult system for business to deal with than outsiders can imagine—to inventory management or, not the least, plant security forces segregation: separate accounting systems, separate plants, and separate engineering and research staffs—at extra expense. Manufacturers do not segregate defense because their commercial lines cannot satisfy military requirements—rather, Gansler told the committee, "because meeting the procurement regulations would seriously jeopardize commercial efficiency and profitability."

Getting defense and commercial operations back together will take more than speech making by the Pentagon or Congress; better it will require true faith in dual use and proselytizing the procurement system in its name. The Air Force has an electronics research center in upstate New York, and Gansler recalled: "We had people coming down from there telling us why you can't do it, why it doesn't make sense, why they have to use Milspecs. There's still that same institutional resistance. And it will continue unless somebody at a senior level in the DOD, with Congress's help, says this is what I want to do." A Defense Science Board summer study in 1990 recommended these changes to Donald Atwood, the Number 2 man in the Pentagon. "Atwood said, yeah, that really makes a lot of sense," Gansler recalled. But Pentagon mills grind slowly.

Yet another coping-with-shrinkage idea is niche manufacturing, or learning how to make money on low-volume production runs.

Streamlined Skunk Works sorts of development, suggested Robert P. Caren, vice president for science and engineering for Lockheed Corporation, where the fenced-off, small design team group was nicknamed. From Lockheed's Skunk Works came its F-117 stealth fighter that earned accolades in the Gulf War, but only about fifty were produced. Contrary to conventional wisdom, Caren believes, small runs can be done inexpensively, allowing experimentation with new ideas. As for no-production runs, developing technology to hold on the shelf until needed, Caren shares the reservations in the defense industry. Technology to him is embodied in teams of designers. Can design teams go on the shelf? Not if they have to earn a living.

Fewer military programs and smaller production pose perplexities of their own. "The toughest one," Caren adds, "is what's to be done about new processes and efficiency. This country doesn't have modern manufacturing methods in composites. Yet these are key to lighter aircraft and spacecraft. There will have to [be] some programs to bring manufacturing technology along so that in the sense it's available if we have to build the defense base back up. Dual use, something the present administration doesn't seem to like, is another way of conserving a lot of this capability." Introducing composites into commercial aircraft eases the transition back into the military.

Enthusiasm or no for dual use, there are opposing views on zero production as an essential defense industry survival tool. Paul H. Richanbach, a Ph.D. researcher at one of the defense think tanks populating Washington, the Institute for Defense Analyses, has been a torch bearer for an alternative: flexible manufacturing.[8] Rather than production runs, Richanbach argues, the military will have to enable contractors to live on research and development. Not enough money has been available even in the Reagan years, to take into production every weapon system that the Pentagon starts, and Richanbach agrees that there will be even less in the future. Hence he doesn't consider the idea of taking a tank or an aircraft out of production, then putting it back on the line ten years later, as necessarily so dumb. Furthermore, he doesn't think it ridiculous to use research and development as an end unto itself to keep design teams together to put such equipment back into production. "Design teams," in his lexicon, does not mean the thousands of engineers drafting rivet holes or hinges in big contractor plants, but the small innovative core, the Skunk Works equivalent.

Processes and methods are part of the picture, too, as Chris Caren noted, and research doesn't help a factory floor learn how to build composite structures right. Using commercial components is not synonymous with dual use, which implies blending potentially difficult-to-integrate defense and commercial technologies. As for using off-the-shelf commercial electronics, Richanbach contends, "we can come up with lots of examples of little things like this where maybe the Defense Department does dumb things, or maybe it doesn't. But that's a far cry from the claim you can have a fully integrated industrial base that provides all defense and commercial needs." His skepticism stems from the aerospace industry, where engines, electronics, hydraulics, and the like are dual use, in both military and commercial aircraft. "The reason for this," he says, "is because these firms have found commercial applications for this defense technology. So the marketplace has worked." Dual use doesn't make technical sense where it hasn't sprouted naturally, in other words.

But, again, a story from Chris Caren, about a Lockheed military airplane for which the government encouraged off-the-shelf—dual-use—technology. So Lockheed found cockpit equipment made by a very good vendor, proposed it, and won. Then the government said, "We just want to change a few requirements." Well, the vendor shot back: "If you want to change those requirements, go and buy from somebody else. We've got a nice commercial line going here and you'll just screw us up." Jet airliners are flying around with the same equipment, but Caren said Lockheed had to follow the more expensive military specification. "The government has its typical bureaucrats, used to doing the job a certain way for a long, long time. As electronics becomes a more dominant cost of the system, we really need to go off-the-shelf."

What all this suggests is that, first, defense companies must buck the system as best they can, working the dual-use, off-the-shelf opportunities to become more like commercial developers. For its part, the Pentagon cannot function as a commercial buyer does, with volumes of regulations and legislation to govern its every step. But it is going to have to think like a commercial buyer. Without a full-spectrum custom industry of its own, it will have to shape its requirements and specifications more to what is on the market or readily adaptable. Instead of designing weapons and forces to need, it will have to design them to cost as well.

Changing the Mix

While full-scale defense company conversion is a recognized impracticality, there is a move toward changing defense business mixes. Dick Linder's strategy at Westinghouse is one case. Lockheed is moving, too, expanding in the airline contract overhaul business and using its electronics experience in a joint venture with AT&T for motor vehicle advisory systems. Another example is Hughes Aircraft. Once it became part of General Motors, Hughes took on the difficult proposition of shifting its share of defense downward and commercial business upward. Traditionally the mix has been around 85 percent defense, 15 percent nondefense until commercial communications satellites whittled defense down to 70 percent. A diversification program is underway, with a goal of reaching 50-50 by end of decade, maintaining as much defense as possible, but going after growth in nondefense areas. Hughes has been running full-page ads trumpeting its commercial gains, not only in satellites but in stereo sound enhancement.

Changing business mix is a perennial in defense. Why defense contractors always pick a 50-50 business mix as a goal is a conundrum. That it is a nice round number that appeals to Wall Street is a good guess. What starts out as diversification tends to end with defense sales falling to the point where the golden 50-50 share is reached—but with a smaller company, as ITT found out. Hughes, with its General Motors connections, is trying an intriguing idea in a legally complex organization, HE Microwave, that amounts to a joint subsidiary of Hughes Aircraft and Delco Electronics Corp, each with 50 percent ownership. Radar is a specialty of a group operation at Hughes. One of radar's future bonanzas might be an active array module, combining a radar transmitter and receiver in one element about the size of a 9-volt battery, or maybe a couple of them laid end to end. They are so atrociously expensive now— using gallium arsenide chips—that even the military can't afford to put 1,500 or 3,000 of them together to equip today's fighter aircraft.

A few years ago Hughes began working under a defense manufacturing technology contract to try to get module costs down to $400 each, which would still make for a half-million- or million-dollar radar antenna. Research also is tied to defense-integrated circuit research—with commercial fallout potential—to try to get manufacturing yields up and costs down on gallium arsenide chips. If commercial applications for the radar transceiver came along, higher production volumes could drive down the costs to perhaps $200 each. At that price General Motors might be able to put one or two modules in an automobile's rear bumper, to look back half a car length or so to make what is referred to pedantically as a near-obstacle detection system. Later, such a system might become a full-blown, front-end road sensor, but now the application is more a mundane annoyance eliminator, backing into a pillar or a neighboring car in a parking garage or running over a child's carelessly abandoned tricycle, for instance. Eventually, if the price drops, a couple more could go into fender sides for obstacle warning in turns.

HE Microwave's crew of 20 was slipped into some empty building space at a Hughes missile plant in Tucson, where drops in military budgets and tough competition left empty quarters. Should research and cost reduction pay off, the detection system probably would start out like the GM's Bose music system or headup instrument displays on windshields or satellite navigation displays, as a top-of-the-line option in high-price cars, and then, if there is market acceptance, as standard equipment on cars like Chewies. Competition is out there, though, from small and maybe faster-moving domestic companies ready for testing bus and recreation vehicle applications and from government-funded European research.

Diversification, United Aircraft found, is not all-purpose insurance. When Harry Gray was brought in as an outsider to shake up the company in the 1970s, United Aircraft turned into United Technologies because Gray bought Otis Elevator and Carrier air conditioning to get into the construction business and Essex to get into automotive wiring. All of its diverse businesses went down together in the 1990–1991 recession. Defense cutbacks came with a slump in construction and automobiles, and United's losses meant it joined the layoff parade.

Metal benders, like General Dynamics, who build ships or airplanes thus have justification for a frosty view of diversification, external diversification that is. Bill Anders calls it a formula for disaster. A McKinsey study for General Dynamics cited an eight-out-of-ten failure rate for acquisitions by defense companies outside of their expertise. That may work for a sinking ship, but Anders says to those who will diversify out of defense: "Good Luck!" What may look like growth will be marginal profitability. Instead, Anders focuses on internal expertise by "deconglomerating." In other words, cash out of business lines that are not adding real value and put the money into better prospects. To the rest of defense, he adds: ". . . Close down and write off truly excess capacity." As part of his strategy to concentrate on four core defense lines, Anders sold the company's missile businesses in the spring of 1992, including the Tomahawk cruise missile that starred in the Gulf War, to Hughes. Doing so consolidated the U.S. missile industrial base in two hands: Hughes and Raytheon.

"People in the industry don't like that," Jack Krings observes. "And obviously the people who work there don't like that. But if he can convert something to cash, before it goes down, it's like selling short in the stock market." To that, add that electronics is an easier transition base than other weapons technologies, and that starting now is too late. Those who succeed will have started commercializing a decade or so ago.

The bottom line? The 1990s shrinkage of defense seems destined to dwarf past episodes. Defense contractors either have to bet the company on surviving in military work or find a way to make a difficult transition into more commercial markets. For the latter, they must wonder what kind of a commercial market is out there for them.

Shrinking into a Tougher Environment

Given that defense will not have a cocoon of its own any more, next must be asked what's out there in the commercial world that the defense industry is supposed to convert into. Technical and professional studies over the last decade have pointed with alarm at the state of military-dominated American research, at the state of American manufacturing, at quality, at corporate management. Shrinking defense may alleviate these problems. By reducing the defense contribution to research, technology, and industrial competence, it may also aggravate them.

Certainly defense is a manufacturing industry—albeit the low-volume antithesis of economies of scale. Nevertheless, defense shares the buffeting that has shaken U.S. manufacturing. Defense shares in the nation's research dilemmas, too, especially in how long it takes to bring technology out of the laboratory onto the production line. Long before the Pentagon policy to stick primarily to research and development rather than production in future weapons development, defense contractors were heavy on research. Why not? Research was funded by the Pentagon, and it was a way into new technologies that would help sell more stuff to the military. Now high-cost companies must either adapt their research and development talents to improving cars or refrigerators or TV sets, or they must gamble on surviving in an uncertain military market.

Commercial companies are more ambivalent toward research, which can be an expensive hobby with uncertain payback. Why they are was put very neatly a few years ago by a veteran government technology manager. Robert A. Frosch has seen both sides.

First in the Navy bureaucracy in the Pentagon and then as administrator of the National Aeronautics and Space Administration, he learned the government ropes. Then he went on to the bureaucracy of General Motors and a research vice presidency. Describing the process of turning research into product, the core of industrial innovation that both commercial and defense industry agonizes over, he observed: "All aspects of the research and technology development precess are creative processes in the artistic sense." Yet the world does give the painter or the poet or the musician a monopoly as artists. As Frosch went on: "Because research is an artistic process, having a research laboratory is rather like owning a stable of poets. People engaged in this work are likely to be individualistic, somewhat 'unreliable' in their rate of production (even in their behavior), and rather variable in their performance. They are likely to impress systematic managers as being extremely untidy and difficult to deal with."[1]

Mastery of this untidy mode of life has meant survival for defense contractors living on what has been at least two-thirds of annual federal research—a $70.8 billion total in 1990. Otherwise, studies at the University of Texas found a narrow research base: that fifty American corporate giants do 80 percent of private R&D. Concentration on research and development has a dark side if it stresses technology at the undue expense of processes. The risk is atrophy of the ability to manufacture, the introduction of new processes on the shop floor. Concurrent with the rise of overseas competition in defense and commercial manufacturing alike over the last 20 years, both defense and nondefense industry have put manufacturing down on priority lists. Where this country once held up the Yankee tinkerer as a model, the designer or the analyst—or both in one package—came to take his place. While this cultural transition was taking place, encouraged by exotic high-tech defense weapons programs and space spectaculars, Japan and Europe began to nibble away at more humdrum industrial markets.

As they did, changes were taking place in U.S. employment. The jump in unemployment from three to five million with the defense cuts after Vietnam was a shock to federal policy makers whose budget planning then was on a catch-phrase assumption of "full employment." They had reason to be surprised, for in the 1960s virtually everyone who could work had a job, at the expense of productivity. As the 1970s unfolded, as overseas competition intensified, as America farmed out manufacturing overseas, defense as well as commercial, that high of five million unemployed became a low. The total slipped up to six million, seven million. Six and a half million out of work was the lowest figure in the 1980s, and the figure hit 10 million in the 1982 recession, eight million in 1991. Unemployment rate highs of 5 percent or 6 percent in the 1950s or 1960s became the norm or the lows in the 1970s and 1980s. Obviously a structural change had taken place in American industry.

Planning and analysis instead of manufacturing became a fixation of American business values, which became expert at it. Or call it by another name, engineering science, as Representative Don Ritter (R-Pa.), a metallurgist by training, did at a hearing. Federal research money goes to academics stressing engineering theory, not to veterans of the shop floor who have spent a career wrestling with processes, and Ritter added a similar note: "We're good at analysis, but you don't get products out of analyzing, even with the biggest IBM computers."[2]

From the empirical evidence of lost market share, big corporate America—and defense is big and bureaucratic—had become mired in size and bureaucracy to the detriment of innovation. A corollary to University of Texas studies was that, the older and larger the Fortune 500 companies get, with some exceptions, the less efficient they are. Yet size has value, as Ralph E. Gomory, former senior vice president for science and technology at IBM, points out. American Telephone and Telegraph's Bell Laboratories housed the creators of the transistor, and IBM's Zurich laboratory similarly sheltered the discoverers of a more recent breakthrough, in finding ceramic components that are superconducting at relatively higher temperatures. Even more recently, IBM researchers have learned how to manipulate individual atoms, with an IBM-developed tunnel microscope. Potentially the technique could mean the ability to produce even smaller electronic components than the marvels of miniaturization routinely manufactured now. Size and innovation have "been very confused," Gomory said. "Small Silicon Valley companies play a tremendous role. Companies like IBM innovate in a different dimension."

While big companies have the cash flow and capital to fund far-out research, they also have the past investment and market position of existing products to protect, a burden the small innovator doesn't have. To give it credit, big industry is not kidding itself about its problem. One of the self-improvement campaigns at the start of the 1990s was to speed up cycle time. Make the transition from laboratory to assembly line faster. Get the technology from the laboratory to the factory line in 12 or 18 months rather than twice that. One of the slowest cycle times has been in defense. There it may take 10 or 15 years to get a new weapon in the field. Government itself caused the slowdown by issuing massive volumes of procedures and regulations for developers to cope with, but industry acquiesced while the money was there. Defense companies have much to learn as the industrial picture changes with military cutbacks.

Detroit and Research

Bob Frosch's counterpart at Ford Motor Company, John P. Mc-Tague, also served in the government mill as a former presidential science adviser. What raises his blood pressure is government's tendency to confuse data with information to make a case that lack of research is at the heart of American industry's troubles. "I know a fair amount about R&D in my own industry—automotive," McTague said. "There are few secrets among the major players in the automotive industry. Ford, GM, Toyota, Nissan, Volkswagen, Fiat. Strangely enough, we each spend about the same fraction of our revenues—about 3 percent—on R&D." Yet of the trade deficit between the United States and Japan, about two-thirds is automotive.

Ratios seesaw in other industries. America and Japan spend about the same percentage in chemical research and development, but the United States has a significant trade surplus in chemicals—without government help. Japan outspends the United States almost two to one in ceramics, and Japan is a world leader. In iron and steel Japan outspends the United States about four to one but U.S. companies are beginning to come back internationally. "That's the good news," McTague said. "The bad news is that they've come back with Japanese technology." Now, in electrical machinery, a rather prosaic-sounding term but whose most important contributor is the computer industry, the U.S. fraction tops Japan; in scientific instruments, it's the U.S. two to one. America at the moment does well in both markets, particularly in computing. For manufacturing overall, U.S. and Japanese companies spend almost identical fractions of revenues on R&D, something McTague notes goes against conventional wisdom. "This trend might have something to do with compensation or costs," he added, "but, taking out ambiguities of dollar and yen, there is an almost identical trend in percentage of employees in [the] company involved in R&D."

Over time, trends have shifted in Japan—down in chemicals, down in iron and steel, up in electrical machinery (read "computers"). Between 1965 and 1975, the trend escalated in motor vehicles. "Five years or seven years before," McTague said sadly, "they really penetrated the U.S. market. Between 1975 and 1985, the same thing occurred in computers. If you don't think IBM is scared, call one of them up." Similarly, in scientific instruments, Japan's relative effort doubled between 1975 and 1985 and U.S. dominance faded. In a macro sense, there is a shifting of R&D effort away from things that are essentially commodities—chemicals, iron, steel—into very high-value, relatively high-technology industries. Whether that's industrial policy, which is the chicken and which is the egg in terms of international competitiveness, is an issue, but McTague added: "I think it can be interpreted that way."

If company-sponsored research and development is comparable, where do the discrepancies come? Emphasis on specific technologies is one obvious answer. Japan does foster a national effort on critical technologies, something the United States is beginning to emulate. Another thing McTague finds appealing is emphasis on manufacturing in general. For the past 20 years at least, manufacturing has been an essentially constant fraction of U.S. gross national product, oscillating around 20 percent, 21 percent, 22 percent. In Britain and France, the figure is about the same, McTague adds, and he asks: "Guess what the percentage is in Germany? About 33 percent. Guess what the percentage is in Japan? About the same. Manufacturing is relatively speaking 50 percent more important to GNP of Germany and Japan, than it is to Britain, U.S., or France. Is it an accident to see similar differences in international trade deficits? I suspect not."

Often heard is the assertion that the United States has entered a postindustrial phase, a service economy as the wave of the future. If so, that is bad news to McTague, who says: "Service industries do almost zero R&D." Using 1985 data, he found service industries spent $35 per employee per year on average for R&D; manufacturing industries $3,500, a factor of 100 difference. Not only is the relative payoff from research and development investment high, but productivity growth in the United States has been two or two and one-half times higher in manufacturing industries than in service industries. "The moral of the story is pretty clear," McTague argues. "As the book says, Manufacturing Matters."[3] Cheap labor offshore or high-value niche market strategies will not erase U.S. manufacturing weaknesses.

An echo to that theme, couched in less catastrophic terms, comes from a former head of research and development in the Pentagon. William J. Perry, who left government after the Carter administration to become a venture capitalist, agrees that the success of manufacturing industry is necessary to national health. But it is not the sole ingredient, at a time of American industry's competitive troubles and the start of a defense dismantling that could affect the rate of national research and development. Quoting Mark Twain on the exaggerated reports of his death, Perry said American manufacturing is alive and well and, at 23 percent of gross national product, is increasing its share. In contrast with an aura that American industry does nothing right, Perry contends: "There is not cause for despair, but I should also say there is no cause for complacency."

Not everything goes right in Japan, either, as in its difficult, but patient, campaign to build a space exploration capability and infrastructure. Nevertheless, Japan's approach to manufacturing puts it first. Japanese engineers spend their first two years on the shop floor, Congressman Ritter pointed out at the House hearing, and that struck a note with one of the witnesses from Boeing, Philip M. Condit, who later was named executive vice president and general manager of the multi-billion-dollar 777 jetliner program that is so vital to that company's commercial aircraft future—and the nation's competitive position in aeronautics. Condit recalled with dismay how one of his engineers had called for shot-peening a small section of a piece whose drawing he sent to the factory floor. In shot-peening, small shotgun-like pellets are fired at a metal part to relieve structural stress. Then that same engineer was asked to leave his computer console and go down to the factory floor to find out what happened at the other end of the flow. There the worker making the part snorted: "Whoever that idiot was who designed this part really made my job tough. I have to mask this entire part so that only this little piece gets shot-peened. I could shot-peen the whole thing, and it would be a piece of cake." So the engineer had to apologize. "That's what I really wanted," he explained, "but I was trying to help." Obviously, the designer needs to understand what goes on in the factory if he really does want to make things easier.

Manufacturing in American industry had become like a faithful wife: taken for granted. In the House Space, Science, and Technology committee's exhaustive record of the problems of American industry, Dr. Roger N. Nagel, professor of manufacturing systems engineering at Lehigh University, used just those words. "We have scripted our children starting with grade school that manufacturing is not necessarily an occupation where you will do well or feel a sense of success as you would maybe in other occupations. And, until recently, there was a notion that academics who either studied or taught manufacturing were those who could not succeed in the other technical engineering disciplines."

Shibboleths like those are changing gradually. More university students are showing interest in manufacturing. Universities are coming on board, too, and Stanford, for example, was selected by the Sloan Foundation for a $3 million gift to create a doctoral degree program in manufacturing. What Professor Nagel says of the academic world has also been true of the industrial world. The best and brightest of a company's engineering talent found a chair in the design office. Those who couldn't make it as designers ended up as production engineers. Not true in Japan. Even in Boeing, Condit said, industrial engineers weren't even included in engineering—by definition.

At the start of the Cold War decades, university science and engineering schools recognized, correctly says Professor John M. Deutch of the Massachusetts Institute of Technology, that analysis and discovery of new knowledge would drive industrial development as graduates and their work filtered into the system. Farsighted as that was at the time, the early 1950s, times changed to the problem of the 1990s, in defense and in the commercial world: how to translate new knowledge into hardware applications. "Take an engineering student at MIT," Professor Deutch said. "Six or seven years after entering as an undergraduate student he would graduate with a doctorate. Never once in those years would he come in contact with a real manufacturing floor or with the real blue-collar labor force in this country."

Academics get preoccupied with the academic track. Without setting foot on the shop floor, students become the next generation of engineering professors. Research and publications are the basis for their evaluation, not execution of a project to build something. Neither is the problem confined to engineering. Tom Murrin, frustrated by the shackles and immovability of the federal government in dealing with industrial problems, left his policy-making job as deputy secretary of commerce in the Bush administration to become dean of the business school at Duquesne University in his long-time Pittsburgh home. Wryly wondering about the business of business schools, he described a survey taken of 280 accredited business schools in this country. Only twenty seven had deans with discrete experience in industry or government. Fundamental attitudes toward such issues as science versus engineering content in departments at universities to adapt to the post–Cold War world will come slowly. "Structural change at a university is not possible today," Professor Deutch said. "Having tried that, I can be categorical." Nevertheless, Murrin points out that Duquesne is introducing total quality management into academe, including benchmarking, comparing its operations within the university and with other schools.

Instead if focusing solely on theory or analysis, Professor Deutch believes, engineering schools will also have to train students to design, to use empirical results. Rather than saluting the brilliant individual star, engineering schools have to train and evaluate students on how they function in a group, in a multidisciplinary environment to make things happen, in process and manufacturing. MIT has a Leaders in Manufacturing program sponsored by corporations where students are encouraged to get masters degrees in both engineering and management. Other universities, Stanford, Lehigh, Rensselaer, Polytechnic Institute among them, are joining the trend toward more real-world, factory-floor kinds of training. Ivy League schools are beginning to see the virtues of what less prestigious ones like the University of Cincinnati and Northeastern found years ago: the work-study program where students spend part of their academic year in industry working, they hope at least, in their chosen field. Business schools are listening, and one survey found more than three-fourths now require manufacturing courses. Next comes the task of getting conservative manufacturers to use their skills.

Another kind of program at MIT is the consortium. One involves the university's military research arm at Lincoln Laboratory with on-campus physics and chemistry departments collaborating with AT&T and IBM in superconductivity research. The corporations pay their own way and the university, luckily Professor Deutch said, found Defense Advanced Research Projects Agency money for its share. Students can spend a semester or more working in an IBM laboratory and, avoiding one of the stumbling blocks of industry consortiums, able to publish their theses. A bit of bad news, though: graduates of the new business-engineering syllabus are being snapped up by investment banks willing to pay them more than manufacturers.

Industry Sees the Problem

As a company once predominantly defense and now predominantly a builder of commercial airliners, Boeing is worth careful study in the context of shrinking military business and in how defense and commercial challenges relate, because it is a company with a foot in each. Like many American companies, defense or commercial, it likes both. Research or manufacturing knowhow from one feeds the other. Its laid-off defense workers could find a new slot in its commercial programs. Besides, aeronautics is the next leading American industry to come under siege from well-financed competitors from abroad.

Engineers wince at Phil Condit's shot-peening story. But they have taken it to heart. Boeing is one of those manufacturers making radical culture changes to restore the engineering-manufacturing balance. Thus Boeing is a microcosm in how American industry as a whole is dealing with the manufacturing challenges from overseas competition. Like automotive companies, they fight the invaders but also are tiptoeing into cooperation and exchange—technology for manufacturing, for example.

Teaming, pioneered in the best-run military weapons programs, is at Boeing under a new mandate of design/build team organization. Historically Boeing would have organized around structures engineers, hydraulics engineers, designers, manufacturing planners, and go on. Each of these functional organizations would have worked in its own area and done its own job. Now, most recently in its 777 airliner program, Boeing has taken all the people who deal with a segment, say the nose section—structural engineers, systems engineers, manufacturing planners, tool designers—and is having them sit together. Since they work together, they don't have to schedule meeting with each other, as they would have before.

Team leaders have a dual reporting line: to their related functional managers and to the program manager. Each team now has an engineering leader and a manufacturing partner. When development is over, manufacturing will become leader and engineering the partner. "It's a new culture for manufacturing," is the way Richard G. Christner put it, "for now we're allowed, quote out of our cage unquote, to go advise engineering. We built this wall between us, a 'them guys' and 'us guys' kind of thing. That's got to go away. We don't know exactly how to do that yet, but we're learning." Part of that learning happens in a new office for Christner, a 35-year engineering veteran of Boeing's heretofore segregated manufacturing, now with new status as part of Boeing's advanced technology organization.

Alternately relishing manufacturing's belated recognition and awed by tampering with a system so steeped in company culture, Christner echoed Phil Condit's shot-peening story: "I've heard people say, 'just give me a drawing and I'll go build it.' Manufacturing didn't want to interfere." Not to mention tooling, which also is reveling in the change.

Tooling for an article the size of a transport airplane, the jigs and fixtures used to hold hundreds of smaller or larger pieces of structure in place while they are riveted or bonded together, was massive enough to be a separate organization. Sam Behar, who is tooling manager for Boeing's new 777 airplane, recalls that tooling, like manufacturing, was a receiver. "When we had this over-the-wall engineering and tossed the design over the fence," he said, "we de-

signed tools to build it that way. Now we're playing a design role."
The new airplane is being designed, its parts fitted and its tooling
created on digital computers before metal is cut. Sam Behar's tooling engineers can say, early enough to do something about it, "Hey,
this part is going to make for a very expensive tool." As Behar
added: "This is the first opportunity we've had to affect cost by
saying, if you design it this way, it's going to be very costly, whereas
if you do it that way we can minimize cost considerably. And the
engineer looks at it and says, 'No reason why I can't.'"

As a rationale for culture change, smallness versus bigness enters.
Dale M. Hougardy, whose 777 territory as vice president is to over-
see operations as a tent covering factory and support functions, is
deeply aware of the issue. "If you look at many small firms," he
said, "even when Boeing was small, that's the way business was
done. The shop guy was here, the engineer was here, too, and they
talked together. As we got large, we got into areas of specialization,
which tended to build walls and draw us apart." Digital computing,
a network system to give the whole organization access to the airplane's design as it takes shape, was the enabling technology, the
glue that put people back together.

This new concept comes together in the 777 program's Digital
Pre-assembly Visibility Center. Formidable as the name sounds, it
applies to a rather ordinary conference room, but which has a computer terminal in one corner where any engineering drawings or
data can be called up and projected onto a big display screen. Blue-
prints aren't used much any more, though there are racks of mylar
sheets along one wall for anyone who needs to look at what passes
for paper drawings these days. And suppliers may want them,
though in Japan they are tapped into the digital computer system,
originally designed in France.

Boeing has been sending people to Japan, some to deal with Japanese companies building parts of Boeing airplanes, some to study
how Japanese industry works. Constant evaluation and improvement are part of the Japanese method, and this is translating into
Boeing's 777. Both Hougardy and Garnet Hizzey, manager of production engineering for the new airplane, talk of lessons learned
from past programs as a formal wicket for design-build teams to
pass through. "We don't have to eat what the engineer creates,"
said Hizzey, who was wearing an American suit but had a distinctly

British accent. Boeing was a big beneficiary of the technical brain drain from Britain to the United States two decades ago that now is in the nascent stage of turning into a brain drain from the United States to Japan. "We can take complex things and make them simple," Hizzey went on. "We can take multiple assemblies and make them out of fewer parts. We can persuade designers to use stuff that's common rather than unique, weird or exotic.

Production flow times won't necessarily accelerate right away with Boeing's new system. They should eventually, but having tooling hand-in-hand with the design phase means the usual tooling manager's lament: "I'm waiting for engineering," should become an endangered species. Japan is good at this sort of thing. Japan's auto companies excel at such concurrency.[4] Putting more people on a project than American companies do in the first year or two, they finish in about three-quarters of the time.

Concurrency teaming is not unique to Boeing by any means. Chrysler is using the same idea under the name simultaneous engineering to develop a new sports car that early on showed great speed in design and development. The gospel is spreading of doing it right the first time in manufacturing, the gospel of concurrent design, engineering, tooling, manufacturing planning, and common databases to help get it right the first time. All that is part of getting cycle time down and reliability up. A disciple is Frank E. Pickering, a veteran of General Electric's aircraft engine business, now vice president and general manager of its engineering division. Typically, he worries, it takes four or five years to get a new commercial engine certified for safety by the government and, unfortunately, perhaps another four or five years for it to meet complete customer expectations. "That cycle time," he says, "is too long."

Like Boeing, General Electric is doing lessons-learned soul searching as a formal process. One sad lesson Pickering recounted comprised four different initial designs for an engine when there should have been one, simply because, in the first place, engineers didn't have a clear idea of the customer and market requirement. In the second place, engineers did not realize technology was not fully in hand to do the job when they did understand it. Besides scrapping the first design, designers had to buy two different sets of tryout parts and wound up taking 30 months instead of what should have been 18 just to get from go-ahead to the first engine running in a test cell, itself only halfway home. Endemic to the defense business, and spilling over into commercial aircraft, is another factor: complex and lengthy contracting procedures in buying components from suppliers.

With concurrency in initial program stages, Pickering said General Electric in its newest commercial engine, the GE 90, lengthened the period for design tradeoff studies. This process of finding how changing one part or one specification affects the rest went from two months to eight months. More time also was spent working with joint supplier teams as a price of avoiding slow subcontracting paperwork. Both were aimed at stabilizing design sooner, using proven designs when possible to avoid typical reinventions of the wheel, having the whole process well though out before manufacturing started. Pickering thinks this new way of managing design can cut a year off product development, enhance quality, trim costs, and wind up with fewer startup problems, the usual failures and fixes for an engine once it goes into service.

Cycle time reduction is a universal theme from American industry scrambling to be more competitive. Using technology to reach this goal has its own pitfalls, however. Raising white-collar productivity, engineering productivity, is one area that R. Noel Longuemare gets involved in as vice president and general manager of the systems development and technology divisions, functional divisions that support both commercial and military project teams at the Westinghouse Electronic Systems Group. "Where's the beef?" is the stock reaction from a project manager. Companies like his have sunk a lot of capital in technologies such as CAD/CAM systems, computerized design, and manufacturing systems that have become so pervasive in high-tech industry. "All these things are supposed to cut down the cost of engineering," Longuemare related, "but our program managers come back all the time and keep saying, 'What happened?'" Their engineering costs are going up, not coming down, because their projects have to carry the depreciation costs of all this investment.

"We are looking at the wrong place for savings," Longuemare answers. "There is a quiet revolution going on, more and more work being done at the front end of systems." With more work at the front end, parts made at different vendor plants nationwide all fit mechanically when they are brought together the first time. And,

Longuemare adds, as an unprecedented change for complex and delicate electronics, they all work the first time. Time is money, and doing the job right at the front end, although it doesn't save costs there, reduces overall cycle time. That in turn reduces overall costs.

IBM learned about cycle time from its own initial flop in the laptop computer market. In bringing a new machine to market in 1991, IBM put together a small team that cut across company boundaries to run the project with less-than-usual second guessing and to take risks in the process. IBM's compartmented, quadruple-check, fail-safe corporate culture had become slow, not nearly as slow as the government, but slow by commercial standards: two or three years to get out a new product. The new laptop came out in 14 months.[5] Defense does have something to contribute here. Lockheed's Skunk Works team approach has been mentioned before, and the IBM experience was Skunk Works in action.

Technologists, being very orderly people, have tended to put everything in little niches. As Westinghouse, like Boeing and General Electric, found, concurrency is a mighty force. Technologies under it are related and synergistic. "We had organized our industries and our entire way of doing business in these stovepipes," Longuemare said, "that tended to be entities in themselves." The growing consensus answer to the problem: multidisciplinary teams, concurrent engineering, do it right at the beginning, faster cycle time. "We've put together small teams to do jobs that by normal standards were considered to be very difficult and expensive," he said. "We've been able to do those jobs in one-third the time. And the costs were even less than that. Management needs to recalibrate expectations, upward."

Motorola found that out in its Six Sigma quality improvement program, which won it the Commerce Department's Baldrige Quality Award. Motorola's self-help effort started at one of the usual sweetness-and-light corporate officers meeting. At the very end, the mood changed when a senior vice president stood up and proceeded to tell the group how sick he was of seeing the junk the company was shipping out to customers. The chief executive officer there listened. From that came exploration into the unknown—and false starts. An initial goal of a ten times improvement in quality in five years was one. "It only took us about six months to realize that was not enough of a reach-out goal," Ray Waddoups recalled. Dr. Ray

O. Waddoups is vice president and director of research and development for Motorola's Government Electronics Group and thus intimately involved in the relationship of technology with a competitive product. "Setting small, incremental goals," he said, "means people will try to get improvements by doing the same thing they have always done, but a little bit better." Shaving three years off that goal reset individual attitudes to: "There's no way I'm going to get there doing what I've been doing. I've got to find a whole new way." From there, the targets became a hundred times improvement in the succeeding two years, and practically perfection by 1992 of six sigma in statistical quality control. "That," Waddoups said, "is 3.4 defects per every million opportunities. We made the first two goals. Right now, we are at about 5.3 sigma, but we are going to fall a little short of six sigma by the target date. This is referred to as successful failure." Did he agree with Longuemare's advice not to shoot too low? "Absolutely."

Processes: An Overlooked Ingredient

Congress, more worried about America's industrial competitive position than the White House, mandated a couple of years ago that agencies put together lists of critical technologies. For the Defense Department, one was done by its advisory industry-government-academia Defense Science Board. Jacques Gansler, who participated, points out something not ordinarily considered in defense: More than half were process, not product, technologies; manufacturing related, that is. "When we talk about technology," he says, "we need to talk about manufacturing technology, not just weapons technology." Microelectronics combines product and process, and supercomputer-aided software engineering, simulation, and modeling, or flexible manufacturing, are precesses. Exotic new materials are not practical without ways to process and engineer them. Defense funding does go for manufacturing technology, but it is a modest share of total military research and development. Neither government nor industry is as eager to fund manufacturing research as they are product, measured by how American industry relaxed on its traditional manufacturing base while overseas competitors, forced by the devastation of war to begin all over again, put more engineering effort into modernization of the factory and its workings.

Perhaps industry takes manufacturing for granted because government does, "perhaps," as Lehigh's Dr. Nagel said, "because there is not a single advocate [for manufacturing] by mandate or by law." If defense manufacturing research is a minor piece of its research pie, it is larger than what the rest of the government has put into this field. Representative Norman Y. Mineta of California, who represents part of Silicon Valley and consequently has taken far more interest in technical issues than most members of Congress, reflected unease at the end of the 1980s over the possibility the United States is putting too much into funding basic research and not enough into applied. He speaks for a growing corps, even in universities, who worry about things such as the $5.5 billion the administration wants to invest in the Superconducting Super Collider to study basic particle physics. "We're not putting our money into manufacturing research and development," he said. "We're putting a lot of money into basic research that's available to others to utilize for commercial applications. . . . Other countries, whether it be Japan, France, Germany or the United Kingdom, don't put their money into basic research. They take the benefits of our basic research, put it into commercial application, into manufacturing R&D. We're left at the dockside so to speak."

Despite the Defense Department's intentions to hold research constant, Mineta recalled that in the cutbacks in the early 1970s, R&D into both product and manufacturing research was pulled back, too. "On the other hand," he said, "the Japanese kept pouring money into research and development on the manufacturing side. By the time the cycle came back in 1976 they hit that street running."

Yet there is a glacial but perceptible change coming in manufacturing research. Not just in companies like Boeing, but also in the Hollings Center approach, in the Commerce Department's National Institute of Science and Technology centers of excellence, and in analogous centers in the more visionary states. And an assistant in the Office of the Secretary of Defense now looks out for manufacturing technology, instead of leaving it just to the services. Faced with massive shrinkage, though, the Pentagon is not a good long-term crutch for manufacturing research.

The Quality War

Closely tied to manufacturing technology is quality. Defense as well as private industry has had its own quality headaches—and still does despite the successes of Desert Storm. For instance, McDonnell Douglas Helicopter in Arizona made not one, but two defense lists in mid-1991: the Contractor Alert List and Contractor Improvement Program, both dealing with quality shortcomings. Paradoxically, McDonnell Douglas has been one of the more enthusiastic—even radical—adopters of the Pentagon's mandated Total Quality Management approach to quality, in its commercial as well as defense operations. For a time the changes raised hob with employee morale at Douglas, where the company's commercial airliners are built. TQMS, the acronym for Total Quality Management System as Douglas calls its version, became an employee joke: Time to Quit and Move to Seattle, where its U.S. rival, Boeing, operates. Despite the initial agonizing, the quality system seems to be cutting down some of the production problems plaguing Douglas.[6]

Where American industry got in trouble, an industrial quality manager will say, was in sinking into complacency in competition-free international markets in the first two postwar decades. In the process, it lost touch with what the customer wanted and where else he could get it. American industry, still in shock at the Japanese invasion, has embraced quality in word and, less certainly, in deed. Total Quality Management was a system—some might call it panacea—that stemmed from Japanese industry culture. Ford Motor Company picked it up, gave it a new name, and one of Ford's executives, who moved into a top Pentagon acquisition job, fostered its adoption in the defense industry. As such systems often do when they are mandated by the government, Total Quality Management became a buzz word for the contractor. It was applied universally, willy-nilly, because the Pentagon said so.

Listening to the gospel at a Total Quality Management workshop one evening was a revealing lesson in how to apply, and not to apply, Total Quality Management. "The minute the government makes it mandatory," one trainee said, "it's dead. Sure, every department will lay out the Total Quality Management plan as it was told. Then the plan will go into the file drawer and that's the last of it to be seen."

Government is the reverse of the Pareto principle, named for the Italian economist's finding that in Italy in the last century, less was more. That is, for example, 20 percent of the population has 80 percent of the wealth. Translated to business that became 20 percent of the staff does 80 percent of the work. Or to quality control, 20 percent of the glitches cause 80 percent of the problems. The principle here is that it takes about the same amount of effort to cut in half the first 80 percent as the last 20 percent; but cutting half of 80 percent fixes 40 percent of the quality problems; but half of 20 percent fixes only 10 percent. As another trainee ruefully noted, the government goes noisily after the 80 percent of the glitches that cause 20 percent of the trouble and never really tackles the big problems in the 80 percent zone where improvement is most needed.

In attention to quality, Japan was there first, embracing the concepts of what crystallized later as Total Quality Management from Americans. Most familiar is the name of Dr. W. Edwards Deming. Deming, brought from America to help rebuild Japanese industry destroyed in World War II, was the classic prophet without honor in his own land until balance-of-trade shock. The reason is that American industry had followed earlier management gurus down the path of assembly lines built on time-and-motion studies, the concept that people there were interchangeable and that inspection at the end would catch the misses.

As Japan learned, and belatedly America recognized, people are not so interchangeable and slogans such as management by objectives are not a substitute for leadership. America saluted the star quarterback and the long pass into the end zone; that is, it shot for the big technical breakthrough to leave the competition behind, granted that the breakthrough might be a bit complicated to design and produce and to work in service. Japan went for the small, simple steps forged by teams, not stars. One small step taken a thousand times makes for a giant step. As management doctrine so successful in Japan, this is called continuous improvement.

American industry's love affair with inspection also went on the rocks. Inspection is not quality control but quality cop-out. Rejects by inspectors are a cost, and worse, specifications for what an inspector passes or flunks allow margins. Those margins are traps. They can let second-class stuff get through to the customer, at least some of which later has to be fixed under warranty. Warranties—especially extended warranties sold at high profit margins—are great sales tools, but are costly to the manufacturer. From a quality engineering standpoint, they are an admission of defeat at the start.

Despite the buzzword effort dust kicked up by Total Quality Management, its principles can work, as these cases show. It will not work by rote, though, and it will not work without trial and error, without adaptation to a specific situation. Oddly enough, one of Deming's fourteen principles calls for eliminating slogans and exhortations. Slogans are an infatuation of defense management especially. Zero Defects was one, and every defense plant 20 years ago sported Zero Defects placards on every wall. Zero Defects was a fine idea: don't accept anything but perfect hardware. Zero Defects posters, though, did little to change business as usual, with reject rates about the same and marginal equipment still going to the customer. In the conference room, slogans sound like leadership. In the factory, the slogans are at best ignored and more likely an irritant, a cause to slow down.

Total Quality Management stresses teamwork breaking down departmental barriers in a company, and it also calls for ad hoc teams assigned to trouble-shoot specific problems. Take the story of what happened when a relatively small division of a big company began to apply Total Quality Management. Allied Signal had moved its AiResearch division making aircraft sensors and air-conditioning systems to Tucson from California in 1987. The move out of congested Los Angeles added disruption to existing travail. Deliveries to both military and commercial customers were chronically late. Inventories were bloated, often with the wrong parts, driving costs grossly out of line. Too much production was assigned to the plant at the start of the year. Then, after running behind schedule for 10 or 11 months, the plant tried to catch up in a burst of overtime at year end.

So first, under Total Quality Management, came the formation of trouble-shooting teams. "Before long," recalled Tim McClung, then the division's quality manager, "we had teams bumping into each other." Furthermore, the teams were picked the wrong way. Too large to begin with, they were selected at the top. As AiResearch learned from experience, teams are best drawn from the group that has the problem and, except perhaps for the leader, picked by those who will do the work.

Eventually, inventories and production schedules came under control. But, as McClung said, the real problem was that the division had lost control of its business. Total Quality Management or any other buzzword is useless until a company understands what is going on internally and how its product meets, or doesn't meet, what its customers want. So the implication is that any simple, logical approach to fix what's broken will work, whether it is called Total Quality Management or something else. And it goes under many names in American industry.

The value of these kinds of exercises is in challenging the conventional wisdom. American industry's conventional wisdom since the mid-1970s is that it is cheaper to build things outside the United States where labor rates are a fraction of those at home. AiResearch had a twin plant in Mexicali, Mexico, a popular way to take advantage of low labor costs south of the border. In its Total Quality Management exercise, the division found that turn-around time to get back from Mexico the assembled parts it shipped from Arizona was five weeks. Furthermore, engineering was in Tucson. So if there was a manufacturing problem in Mexico, the workers solved it their own way, for better or for worse. Solution: close the Mexican plant and pull the work back to the United States. Turn-around time dropped to one week, the liaison with engineering saved much in quality grief, and the new operation was cheaper. If Total Quality Management only forced a reappraisal of overseas manufacture, the system would be worth the price of admission alone, purely in terms of control over manufacture and quality. Despite the slogans quality as a U.S. industrial goal was slow in maturing. A survey sponsored by the American Quality Foundation in 1992 found U.S. companies paid less attention to customer complaints than their overseas competitors and quality running far behind profitability as a measure of a senior manager performance.

Can Defense Help or Hurt?

Where defense can provide principles like small, dedicated Skunk Works teams, its practices usually are not much good for commercial companies to emulate. Contractors are used to building their profits on a cost basis accepted by the government, not fitting technology into a budget. They are used to throwing money at technical obstacles, not working around them to keep prices down.

Dick Christner of Boeing has a nice capsule description of the difference between military contracting and commercial: Military contractors seem to have more freedom in pricing. If the Air Force says, "I want you to add another cannon on the wing," the contractor can reply, "Gee whiz, that's a whole new basis for costing. That's going to cost you a zillion dollars." Commercial work doesn't get that freedom. If Boeing is building an airplane and United Airlines comes in and says, "I want you to add another telephone for the stewardess, but I want it hanging over here, not over there." Is that a new basis for costing? No. "It's free!" Christner said. "We don't get to do it for a new price."

Ford's John McTague has his own perspective in looking at what he calls two 800-pound gorillas: the aerospace and automotive industries in the $100 to $200 billion annual sales bracket. While it is the commercial wing of aerospace that generates a $26 billion annual trade surplus that McTague was comparing to automotive's $41 billion deficit, commercial aerospace has picked up so many habits from its military cousins that it is also a good proxy for comparing commercial versus defense habits. Operating scales are vastly different, aerospace producing 2,500 commercial aircraft against automotive's 10 million annual units, but using twice as many employees to do that. Adding military and space vehicles and even small missiles would not get aerospace close to a hundredth of automotive volume. To be fair, though, aerospace—military especially—is a subsystem and support equipment game, subsystems that are precision units themselves built in larger volume; engines, electronic black boxes, displays and instrumentation, solar panels, or prosaic things like landing gear struts, wheels and brakes, heavier or light and delicate, more complex and more expensive than even a luxury. Revenues per employee in the automotive industry are far higher than in aerospace, which is a very low-volume, very high-value-added industry.

"Is there any way these two industries can do things together," McTague wondered. There are so-called generic technologies in common, such as electronics. Now approaching $2,000 per vehicle, electronics value added includes a typical fifteen microprocessors. Sensors are a related technology for obstacle detection or guidance, and so are navigation systems. Light, reliable materials are another, to meet a critical need to take weight out of vehicles to meet federal fuel mileage standards. Computing is becoming more important to automotive industry in terms of design, engineering, crash simulation. Aircraft manufacturers do more of this on the ground before expensive flight testing; automotive companies have been able to afford cut-and-try on the test track, but those times may be changing. Ford hired aerodynamicists out of the aerospace industry to design its then precedent-breaking, streamlined Taurus body shape.

"Maybe we can get some synergy out of all this, and at the same time some from that one-third of national research and development that comes from the Defense Department," McTague suggested. "But it's not so easy." Value per pound of typical airplane is $300; per pound of Taurus, about $4, including air conditioning, microelectronics, engine, and so on. In one year the auto industry uses 23 million tons of steel, plastic, and the like; commercial aircraft use 160,000 tons. Very different criteria, even with common interests.

"If we are going to get synergy out of these so-called generic technologies," McTague said, "and I believe we are, we've got to address this issue up front." Research on, say, structures and materials is not enough. From the beginning, processing for high volume, for low cost, for reproducibility without having to do extensive hands-on testing must be addressed as well.

Opportunity for cross-feed among these industries does exist, as McTague suggests. Defense and defense companies can contribute. Still, peace is going to be tough for defense companies and defense technologies in an increasingly commercial world. There is a narrow bridge, though, in dual-use blending of commercial parts in military hardware.

5

Technology Transfer: Another Way

Beating plowshares into swords is easier than beating swords into plowshares. Hysteresis, as engineers call it. What goes up comes down—but with a lag, a different shape to the curve. That's why commercial companies have an easier time getting into defense contracting than defense contractors do diversifying out. There is an alternative, though, to outright conversion of defense contractors, as alternative for the nation as a whole to wring a return from investment in weapons development: transferring defense technology for others to use. Despite misses, Ford's John McTague's synergy has hit pay dirt, too.

Government could help the process along, former White House science adviser William Graham believes, if it would think early enough about dual-use technology. A physicist turned civil servant as NASA's head during the space shuttle Challenger's disaster prior to his White House tour, Bill Graham has again turned small-company entrepreneur. "As far as I can tell," he said, reflecting on the history of the Air Force jet tanker that became the first American jet airliner, "as a policy base, it's analogous to the fact that even a blind squirrel will find an acorn once in a while."

Once the Air Force built high-speed jet bombers, it found they were great but guzzlers of fuel. So the hunt for the acorn started with airborne tankers. These initially were converted, propeller-driven B-29s, that turned out to be too slow to support the faster jets very well. To deal with that, the Air Force hired Boeing to design and build a jet tanker.

"They made one decision together," Graham said, "one key technical change, one big step which has led to Boeing's supremacy in the commercial aviation field." Instead of the easy way, building the KC-135 with only the cockpit and tail gunner's compartment pressurized, as the B-47 and the B-52 bombers had, the developers took a longer view. "They pressurized the whole damn fuselage," Graham said. "Now they were just a small increment away from a viable commercial jet transport."

Not only did that early foresight lead to a family of commercial airplanes, but it also was like bread cast on the waters that came back as a ham sandwich. All of the commercial airplanes that the military uses today—the E-3 warning and control version of the 707, the E-4 presidential command post version of the 747, and the KC-10 tanker version of the DC-10 among them—were developed originally as commercial products, essentially free for the military, in Graham's opinion.

Costs were less because these were part of a major commercial production operation. So the military got its money back many times over, though it could have decided: "No, we know more about unpressurized hull planes. Let's build the tanker that way." This is what Graham calls a vernier kind of decision, a relatively small adjustment that led to America's biggest high-tech export: over $38 billion in good years. "If we're going to do something for national security anyway," Graham said, "then we should look very carefully at possible commercial applications of that technology. With very modest, fine-tuning vernier control on military programs, it's possible to develop a commercial industrial base as well."

Another possibility Graham likes is a tritium-producing reactor the military and Department of Energy have been looking for to replace the geriatric lineup at Savannah River that has none of the modern understanding of safety. Like an unpressurized jet tanker, the most obvious choice was a newer incarnation of the current Savannah River reactors. "They are heavy water moderated," Graham said, "and run at boiling temperature and atmospheric pressure. They have enormous energy output, several thousand megawatts, but it's all low-grade heat. There's no way to produce power from that."

As White House science advisor, Graham encouraged the Energy Department to follow twin, dual-use tracks. One would be based on the existing Savannah River technology, substantially redone to bring it up to modern safety standards. The other was to consider a tritium-producing reactor that would establish the technology for a new approach to commercial power production reactors.

"If this were Japan's Ministry of International Trade and Industry sitting around talking about these things," Graham speculated, "MITI would be saying: 'Look, we've got this reactor which is going to be the basis of a power reactor industry. By the way, isn't there some way to figure out how to make tritium in it, too?'" America was doing the reverse. A better, intermediate position would be to give some weight to dual use of the technology, Graham said, adding: "We're not going to let it overwhelm the national security use. But, like the KC-135 tanker's relatively modest investment in pressurized hull technology, why not let the taxpayer get some return on his investment?"

The Department of Energy did decide to pursue two parallel paths. One was the sure bet through all of the safety and environmental analyses; the second was a modular high-temperature gas reactor built of refractory materials that produces very high-temperature helium as its working fluid, for power as well as process heat, much higher than a pressurized water reactor. Small enough to dissipate residual decay product heat in a shutdown without needing further cooling, it avoids the reactor's damaging itself. Graham believes that U.S. reactors, with big containment structures, aren't going to hurt anybody else, unlike those from the former Soviet Union, but they can hurt themselves. "Companies," Graham said, " are just not enthusiastic about having $3 billion assets converted to $3 billion liabilities in ten minutes. So we don't find a lot of private capital going into new reactors of the present generation. If we could say with confidence a new reactor wouldn't melt itself down, that would be a more commercially viable approach to nuclear power." With the Soviet collapse, though, the new tritium reactor may become academic. A go-ahead has been delayed until 1993 for sorting out the question of whether the United States needs more tritium for weapons.

The Conversion Blues

For every Boeing 707 jet transport or its Pratt & Whitney J57 engine that fathered a similar family of commercial power plants, there have been defense technology transfer ideas that simply vanished. A wonder glue that bonded the plywood in the famous World War II Spruce Goose looked like a sure winner in the postwar commercial market. It wasn't. Some estimates place defense product transitions to commercial markets as low as 1 to 2 percent of sales; others put them as high as 10 percent, although there is some confusion over whether sales to commercial airlines are included. Nevertheless, the list of abandoned commercial ventures is long, ranging from stainless steel caskets to powered wheelbarrows to garbage reduction machinery.[1] Nevertheless, it is easy to confuse defense company commercial diversification attempts with transfer of defense-developed technology.

For the former, there is Grumman, an old-line Long Island producer of Navy fighters. Grumman's foray into the bus business with the acquisition of Flxible, an experienced manufacturer in this field, ended with embarrassment. Flxible's busses for New York City broke down as much as they ran. In this case, defense management was not able to straighten out a company already in trouble any better than nondefense management. But Grumman was able to transfer aircraft aluminum manufacturing experience into canoe building and truck bodies, the latter in a long-term contract for building delivery trucks for the U.S. Postal Service. More often the transfer of defense technology to commercial companies works better, as in disk brakes and antilock braking systems. Now standard equipment on automobiles, they were developed originally for military and commercial aircraft. More pervasive in society is the computer, initially funded with defense money and developed with defense as a customer. To the surprise of its early practitioners, the commercial market exploded eventually to dwarf defense.[2]

A recent example is Motorola's Micro-TAC Lite cellular telephone, beating the Japanese into the market by about 18 months. Essentially a radio transceiver about the size of a calculator, Motorola with a lead could price at a profitable $3,700 as the sole entrant. Cycle time is indeed vital, for Japan has now fielded its own phone, and Motorola's price has dropped—it is advertised as low as $800, profitable but thinly. Micro-TAC Lite, about half the size of the first version and with battery life three times longer, has since hit the market, list-priced at $1,200. How much Micro-TAC had roots in defense is a yes-and-no situation, answers Ray Waddoups,

Motorola's Government Electronics Group research vice president, with combinations and iterations along the way. Micro-TAC Lite did come more directly out of defense, in that a smaller defense organization not part of the Government Electronics Group did most of the development.

Though there are contentions that the rate of technology fallout from the military has dropped off in recent years, Ray Waddoups says that Motorola has done more with its defense technology in the last two or three years than it had earlier. Iridium came out of Motorola's defense group, a commercial satellite cellular radio system venture to break away from dependence on and investment in ground stations and on urban center customers. Third World countries with no communications infrastructure could take a short cut to a telephone system without the cost of stringing wires and building switching centers. As a technically creative name for the prospective corporation, Iridium stemmed from the satellite constellation of seventy-seven private spacecraft, the same as the number of electrons in the element.

The Communications Revolution

As with a Micro-TAC, the transfer of communications satellite technology from the military was anything but clear cut. Not only was the space agency involved, but so also was Comsat, a privately funded company created by the government in the Kennedy era to exploit a then little-tested technology. But as retired Navy Admiral Bobby Inman says from a perspective as former director of the U.S. National Security Agency, the communications eavesdropping capital of the world, the United States got its lead in communications satellites because of large government investment.

From a vantage point in the world of reconnaissance, Albert D. Wheelon watched the story unfold. Later, as an industrialist, he traced the military and commercial relationships.[3] Ballistic missiles were the core of the military contribution. "In a classic example," he said, "of swords into plowshares, these rockets provided the practical road to peaceful applications of space. For more than 20 years, the ancient warriors Thor and Atlas have been launching communication satellites."

In 1957, Sputnik, the Soviet pioneer spacecraft, did as much as anything to galvanize a few theoretical, science-fiction-originated ideas for communications spacecraft into urgent development. "The military took a bold lead," Wheelon wrote, "initiating the Advent program in 1960 to place a communications satellite in synchronous orbit." Synchronous satellites were indeed a bold concept then, for they needed breakthroughs in launcher thrust and in spacecraft guidance and control to put them in a 22,000-mile-high orbit at a precise latitude and keep them at an unchanging station over the earth. Two years and $170 million later, big dollars in those days, the Defense Department stopped the work. Embarrassed by failure, the military turned cautious. While AT&T touted a network of satellites in low orbits, Defense Department leaders turned doubtful.

To a visionary-sounding claim in early 1962 from Allen Puckett, then a Hughes vice president, that an advanced version of the company's Syncom satellite could provide a commercial communications system by the middle of 1963, Secretary of Defense Robert McNamara replied in August: "I think he is unduly optimistic. . . ." The head of the space agency's communications satellite work added that an operational synchronous system was not within the state of the art at the time. Harold Brown, later defense secretary himself but director of Defense Research and Engineering in early 1963, conceded that synchronous satellites might be the ultimate system, "but they are probably many years further on than a good medium [altitude] system." AT&T, with a massive ground infrastructure investment, pushed for less capable nonsynchronous systems in a classic case of the conflict between innovation and protecting the status quo.

Advent had a cauterizing effect on the military, Wheelon thinks in hindsight. "All this is second hand, because I didn't get directly involved until later," says Wheelon, now retired. "But there were a lot of screwups. My understanding is that the Advent satellite had weight growth and the Centaur launch vehicle had payload degrowth. Problem number two is that they were trying to use triodes [a kind of electron control valve] in the satellite's electronics and you couldn't hardly get from here to there with triodes." Syncom's later success was built largely on work by Bell Laboratories in induction-cooled traveling wave tubes (far better than triode tubes for signal amplification over broad frequency bands) and by a Hughes engineer who figured out how to make them lightweight.

Nevertheless, Defense and NASA, more or less as insurance if Advent got into deeper trouble, sponsored a flight test of Syncom, which successfully reached synchronous orbit less than a year later. This was fortuitous, for Hughes had used its government-contract-shared independent research and development money to develop Syncom and it didn't have the funds or facilities to go into flight test itself. While AT&T did put its own medium-altitude satellites in orbit, the government's creature Comsat faced a controversial choice in which direction to take an American entry into what the Kennedy administration envisioned with prescience as an international communications satellite cooperative. McNamara proposed a joint military-civil system, which would have a big defense customer at once to make for good economics. Civil users objected that the military would demand jamming protection, radiation hardening, and other things that they could not afford. Eventually the joint system collapsed under congressional scrutiny. Despite early lobbying by Defense, NASA, and AT&T, Wheelon continued, Syncom's success convinced Comsat to opt for that technology, which dominates the field today.

Technology transfer from the military, as communications satellites show, is neither the fiasco portrayed by defense critics nor an easy way out. The real process is convoluted, as in this case, and difficult to trace or fathom. Normally the military advances technology simply by providing an instant customer base. Not always, as this instance showed. Normally the government's skepticism about promises from technology salesmen is well founded. In this instance, the skeptics were too conservative. At the core is simple obviousness: it takes vision and evangelism at the right time, but hard-headedness, too, for sensing when the time is wrong. As Bud Wheelon puts it now: "It's a kind of sensing those natural intersections of what you can do versus what you want to do."

In Search of the Grail

Frank Moore talks like an Air Force fighter jock. Yet he was a desk-flying commander of the Avionics Laboratory at Wright-Patterson

Air Force Base in Dayton, Ohio, where U.S. aeronautical technology began, when he retired to do something else. Now as non-GI Ernest F. Moore, engineer, he is director of the Edison Materials Technology Center, or EMTEC, one of a family of research centers the state of Ohio is using as solvents to the miseries of the rust belt. He is walking-around support of the thesis that transfer of military technology is lagging behind because it has not been pursued hard enough.

Congress is willing to listen, because it has enacted two laws, the latest the Federal Technology Transfer Act of 1986. Both lay on government laboratories the duty to get likely technology into the commercial stream, although the newer act is stronger. Whether technology transfer can be forced is an uncertainty the law may clarify, but in the meantime it has made something perfectly clear. Pursuit of commercially applicable technology in government laboratories is now legally okay. While it still could be career limiting, commercializing technology won't send any civil servant to jail. The wall is coming down.

"I think the wall existed because of people like me," Moore admitted. "When I was inside the Air Force, I did not have a mission to do technology transfer. "'Your job is to fly and fight,' I was told by the military commander, 'and don't you forget it.'" Anything outside that central mission was suspect, a misuse of resources, unfocused. "If someone walked in off the street back in those days and asked for technical help," Moore said, "he'd probably get the runaround."

But the mission now has changed. Congress is asking what all these federal agencies and laboratories are doing to help the United States in an era of intensifying technical and industrial competition. In Moore's opinion, it has been idiotic not to capitalize commercially before now on the investment made in defense technology.

Moore's state-sponsored office is in a grassy, tree-dotted research park newly sprung up in Kettering, conveniently near the historic Wright Field complex near Dayton where the Wright brothers and Air Force research got their start. "We have the only memorandum of understanding of its kind," he said, "which allows me to use all the facilities and people at Wright Patterson Air Force Base cost free." This is a culmination of what used to bother Moore when he was on the inside at Wright Field: watching all those marvelous gadgets coming out and wondering why they weren't finding their way into commercial markets. "If I ever get out," Moore thought to himself, "I'm going to figure out a way to improve that process."

What Moore's research outfit has learned is to play to small and medium businesses that don't have the resources to dedicate consistently to technology, those who can most use outside help, even from the government. Problems they are facing may have been solved a decade ago by the Air Force or somebody else. "To get them in and start the transfer process," Moore said, "we created a thing called Request for Help. Any member can walk in off the street with a problem in our technical area, which is industrial materials and end processes. For a stopped production line or a busted part, we literally respond in near real time. Forty percent of the time, the response has been finding an Air Force technology that was on the shelf."

To find technology that fits into the right slot has been a chronic shortfall in research and development, military or otherwise. Moore's center, one of eight ranging from welding to biotechnology that was started in a spasm of foresight about a decade ago by the state, uses a computerized database with a network of people skills—whether an expert is a mechanical or electrical engineer, whether he's an expert on chrome rods—as well as technologies, facilities, and equipment.

Of the materials center's industry members, some are large companies with Ohio plants, like General Motors, General Electric, Du Pont, NCR, and Mead. These are important because they can contribute the technical horsepower for projects. "But we have members all the way down to small businesses like Brinkman Tool & Die, Inc., where Chuck Brinkman himself signs the checks," Moore said. There are a dozen universities and several research institutes and federal entities. But small and medium-sized companies are essential political factors, as well as the place where technology transfer can be applied. As Moore recognized:

If you can't show a small or medium company guy that you're interested in his cliff, his two-to-three week away cliff, what he looks at as survival, if you can't help him make it through the next quarter, if you can't do something for him in real time, you will probably not get his confidence, ever. By EMTEC's doing these requests for help, maybe he has a fix in one hour. He brings a stack of broken parts in here, and his production line is stopped. And he lays them on the table and says, "I'm a member of EMTEC, do something." If we solve that, we have an ally for life.

Core research projects at the center are shop-floor stuff, beamed at this kind of membership, carried out by teams from members, and two-thirds of which as proposals don't make it through the selection process. A random example: a study of high-temperature creep of titanium aluminides. A task leader for this project was volunteered by General Electric's aircraft engine plant near Cincinnati and the group had participants from the University of Cincinnati and the Air Force (a participant in more than half the projects), as well as Armco Steel and the Duriron Company, all of whose in-kind participation was worth $210,000. Some are in more exciting electronic chip–making technologies, such as stereolithography, with over a dozen participants and a million dollars in time and cash. Others are on a smaller scale. As Moore related: "How do you cast a 42,000 lb. piston? Here was a guy who was scared to death, who hadn't done anything this large before, who didn't want the risk of pouring twenty tons of molten metal only to find out later that he had poured a lot of porosity into it. Then he'd have to melt it down and start over." Using software that had been partly developed by the Air Force, but then modified and improved and applied to this problem, his casting was redesigned five times in three days on the computer. Then, on the computer, the team watched an electronic version of the casting solidify while the pores were weeded out.

Of the consortium's sixty-plus participants, the big company member pays $50,000 a year to belong, the small company $750. Moore is adamant about having no staff, or not one to speak of. "When you form a technical staff, you have mouths to feed. If I hire a metallurgist, and a guy comes in with a problem that clearly has a ceramic answer, or a plastics answer, I'm going to tell him it's metal. I'm going to put Joe, the metallurgist, back to work." So the membership is staff.

Academia is a valued asset—sometimes. Weathered, self-made entrepreneurs, for one, do not relate very well to esoteric research or to college professors. Take a joint industry-academic team to an entrepreneur who has survived somehow, Moore finds, and he'll sort out academicians in a minute. And he'll put them in what he thinks is their place. For his part, Moore gets steamed up about academic research: "University guys clone themselves five to one," he snorted. "They're looking for an academic research career, that's all they know how to do." While the country talks about not investing enough in R&D, Moore contends the investment is larger than ever. "Clones," he said, "are growing in geometric progression, but the budget's only growing linearly. So, sure, there's a shortage of R&D money. The reason is there's an excess of people hunting it." Industry loses patience with academia's developing a lot of theory for which there is no application. EMTEC is sensitive on that point. "I just terminated one researcher just this week at Ohio State because he was trying to do that," Moore said. "I couldn't convince him that he needed to do something useful."

Abstractly searching for knowledge in science is one thing. Engineering is another discipline, Moore argues—the discipline of applying that knowledge to solve problems and create products. Abrading away that distinction figures in what has gone wrong in American industry. "Even when I got my doctorate years ago," he said, "they told me I had to go back and do original research. I did, solving five problems the world had never solved. Why didn't they tell me to put together these fuzzy ideas and show how I would develop a product and market it and get it out?" This is the classic conflict between engineers and scientists, academics and industrialists.

"Nowadays," Moore warmed to the issue, "there are kids in their 20s and 30s, who graduate as mechanical engineers, and who wouldn't know a five-axis milling machine if they fell on it while it was running." Once Moore saw an entire model of a factory at a briefing on integrated manufacturing in which personal computers were everything, the casting operation, forging, stamping. "Do you know what it is you're modeling in that PC in the casting process," Moore asked. "Oh no," the fellow doing the briefing answered, "we'll develop that later." To Moore that meant he didn't understand the processes he was going to deal with. Generalize that, and American industry's manufacturing travail grows clearer.

For military technology transfer, with a new regional technology center, Ohio is going to have its people with an industrial perspective residing at Wright-Patterson, collocated with government scientists, watching the technology develop, ready to pounce on likely-looking innovations. Two of Moore's people will go at least part time to the Air Force materials laboratory. "If," Moore speculated, "there had been someone like that in the days when the Air Force wrote the radar specification that read, 'Don't stand in front of this thing, it will fry your leg,' he would have said: 'Hey wait a minute, if that will fry my leg it will fry bacon. Why don't I figure out how to make a microwave oven out of this?'" That's the type of person Moore's center wants scrounging for technology on the Air Force shelf.

"Our philosophy is that the game has changed," he said. "People (i.e., Japan and Europe or any other state) are figuring out how to put together their best teams and to field those teams in academia, government, and industry. Unless we learn how to play this new game, we're going to lose our butts." Ohio needs to think about Wright Patterson's role in its strategic economic development, just as New Mexico or California needs to worry about what might be secreted at the nuclear weapons laboratories there.

Of course, Japan and Europe, but other states? Indeed, especially Texas, where technology venturing has become a label for a combination of technology transfer and new company starts. George Kozmetsky has been instrumental in shaping this strategy, which integrates state or local funding, but not excluding federal; venture capital from wealthy Texans as business angels, not the usual venture capital firms Kozmetsky thinks have fallen into hands of amateurs; and university and industry business and technical talent. All this Kozmetsky and his colleagues call a smart infrastructure, essential to cope with the new "soft" Cold War. In it, states or regions will outpace the federal government in evolving their own technoindustrial policies tuned to traditional local economic expansion.

Kozmetsky shared in the founding of the defense/commercial conglomerate Teledyne, he says, so he could retire to run a business school. Eventually the University of Texas at Austin liked his ideas. Since, through the IC^2 Institute, a research center for innovation, creativity, and capital he founded, Kozmetsky has sponsored, among others, forums and books and has written articles himself dealing with technology utilization, especially commercializing defense technology, and new business incubation. It is no coincidence that in its enthusiastic climate for technology, Austin landed facilities of the new electronics research consortiums Microelectronics and Computer Corporation and Sematech. To grow spinoffs from these new age research consortiums, the city of Austin put up $50,000, the Chamber of Commerce $25,000, and private sources $50,000 to form the Austin Technology Incubator with the University business school and the IC^2 Institute. Its networking circle is 2,000 professionals from business and industry who commit 100 free hours a year to advise new companies. Then there is the Texas Capital Network, the business angels, 100 of whom are interested in investing $40 million in startups. Beyond that, Kozmetsky said, is $68 million in state funds that technology entrepreneurs can compete for.

Important as federal research and development money is, Kozmetsky's experience breeds impatience over federal technology transfer catechism. Huge data banks are one illusion, as if answers will just pop out of a computer. "Every time anyone gets serious about technology transfer," he said, "it's person to person."

Pushing the String

Congress did four or five clever things in writing both technology transfer laws. In the first, the Stevenson-Wydler Technology Innovation Act of 1980, Congress created special conduits to disperse technologies from federal laboratories, called Offices of Research and Technology Applications, or ORTAs. Until the defense budget shrinkage became more than a temporary phenomenon, the office at Wright-Patterson had the perfunctory one full-time employee. Then, as the impact of the second law, the Federal Technology Transfer Act of 1986, began to widen, as the need for laboratories to reinforce their usefulness became more essential, it grew to five people. With the larger staff, says Colonel Richard R. Paul, who commands the Air Force Wright Laboratory, his group intends to become more aggressive in marketing technologies to the outside world.

Not only did the later law reverse the past and make technology transfer a requirement for government scientists and engineers, it also did something even more persuasive. Government innovators can now share in patents and royalties for work that gets into the commercial marketplace—up to $100,000 per case. Furthermore, the law blessed the formation of cooperative research and development agreements, known by the acronym CRDAs, between government and private industry. Where laboratory directors in the past might have shied off working with private companies or, in effect, putting federal dollars to the benefit of a local community, the new law encourages both. In fact, the federal government contributes time and materials rather than cash, as in its programs with Frank Moore's Ohio materials research center.

Not only was the technology transfer act important, but the head of Sandia National Laboratories, Albert Narath, adds that more legislation along the critical technologies line is possible. This would go after leveraging more technology transfer by permitting partnerships of private corporations, national laboratories, and universities to develop critical technologies. Such lists are sprouting up all over the federal government, partly in response to congressional prodding. "Technology transfer is largely a misnomer, or at least that's a prejudice of mine," Narath says. "What we need is to focus on technology development in a teaming arrangement. Very seldom does a national laboratory have something sitting on its shelf ready to be dusted off and moved into the marketplace." Like the Ohio experience, the Wright Field experience, government laboratories, Narath believes, would do well to assemble advisory teams from industry to help improve their understanding of industry's technology needs.

Sandia, Livermore, and Los Alamos, unlike Wright Field and the military laboratories, are GOCO, translated as government owned, contractor operated—in this case government being the Department of Energy. So their legislative and contractual charters are different. These institutions did not get approval from Congress to join into cooperative research agreements with industry until 1989, three years after the military laboratories. (These CRDAs are phonetically "craydas" in federalese.) The really vital change, says Olen Thompson, department manager for technology transfer applications at Sandia, was in exclusivity. Unclassified work at government laboratories, once open to anyone, can be sheltered under a cooperative agreement for five years against Freedom of Information Act prying by competitors looking for cheap shortcuts. "No industrial company is going to invest several hundred million dollars in a government laboratory technology if everything is public information," Thompson said. "If the company can't get a head start in the market in five years, then the technology probably wasn't worth much to start with in today's competitive world."

By the time Sandia's operating contract was amended to allow cooperative agreements, two years had passed. To some amazement at the laboratory, there were eight approved agreements by the fall of 1991 and forty-nine more in the pipeline, a much faster buildup than Thompson or his colleagues expected. What makes for cumbersome agreement negotiations is the legislation's rules of the game. Technology transfer must benefit U.S. companies, but in today's buy-out and global ownership maze, it is harder and harder to decide what is a U.S. company. "That will always be a nebulous issue," Thompson said, "and can get real complicated." But the best yardstick seems to be whether manufacturing of the ultimate product is done in the United States.

Another legal issue is fairness. While exclusivity is permissible, if not encouraged, the laboratory must offer opportunity to everyone. So laboratories publish their work in technical journals as a way of advertising. They also use the *Commerce Business Daily,* a government newspaper-style bulletin little known outside the defense contracting community. It consists of what resemble want ads. In reality, they are military and civil agency requests for proposals or advertisements of other kinds of opportunities to do business with Uncle Sam. If more than one company is interested in a cooperative agreement, then there may be a run-off. Sandia will invite interested companies to a briefing. "If three or four companies want to pursue the technology," Thompson said, "we might suggest they combine resources, work with us to bring the technology to a certain level, then compete from there." Precompetitive research, in other words. "If that doesn't work, we may have to make a judgment that this technology could only be effectively commercialized by one company, or choose which one had the best plan." Hence Al Narath's comment that all the talk about picking winners and losers is a bit irrelevant. Fairness, the test of reasonableness, becomes the issue.

Small business has learned the hard way about how difficult it is to deal with the federal government, and how slow. Delays have meant coming to market six months late, and exclusivity and patent protection still bear on the process, and the Energy Department facilities did need catch-up legislation. Royalties to federal laboratories as a result, amounted to just $9.4 million in 1990.[4]

Technology has moved out of military programs at Sandia. Drill bit improvements in the commercial marketplace stem from its energy work, as in the drilling in underground nuclear testing. Thompson says Sandia fathered laminar flow clean room technology that has spread extensively into microelectronics and pharmaceutical processing. Still, Thompson says, government laboratories have a lot to learn about speed and certainty in dealing with industry. "Accomplishing that in a bureaucratic world is maybe a little counterculture," he concedes. "We've already had people walk away, people who saw their market window go past while we waited for legislation or contract modifications"—to allow exclusivity.

Another Energy Department laboratory director, Siegfried S. Hecker, at Los Alamos, likes the collaborative approach with industry, pilot centers with cost sharing where no money changes hands. Government funds the laboratory, industry its own work, with objectives set jointly, and with technology not pushed but pulled by industry interest. Three such pilot centers in their two-year existence have signed fifty-six agreements with forty different companies. Los Alamos is working with a materials company, Du Pont, and a device company, Hewlett-Packard, on high-temperature superconductor microwave applications—far-out research that is beginning to show potential for product. Equally far out are particle, laser, or ion beam technologies from the exotic world of Star Wars, the Strategic Defense Initiative. Los Alamos is trying to take its defense work on these into potential commercial areas. In particular, it is researching free electron lasers to increase power, efficiency, brightness, compactness, and reliability. Work has reached the point where the free electron laser, Hecker says, may become the ideal light source for ultrashort-wavelength photolithography with resolutions of 0.1 micron or even 0.05 micron in terms of light source. With Texas Instruments, Motorola, and, to some extent, Intel, to draw on all the rest of photolithography system technology, collaboration looks to leveraging federal technology investment for commercial fallout in electronic chip manufacture. In Hecker's view, Sandia is doing very interesting work with a consortium in specialty metals processing. Lawrence Livermore is working with industry in superplastic steel technology.

Government laboratory technologists generally agree to a fundamental in this kind of commercial application research: it must be in the laboratory's core competence areas. Then the technology, Hecker says, "must benefit the customer, not the federal agency." Finally, from the purely practical standpoint, government contracting procedures are to slow and too encumbered with complexity to deal in the commercial world. On industry's side, cost sharing is equally important to make sure there is a real commitment by industry. Nevertheless, government-industry cooperative agreements grew about 40 percent in the two years after 1990, to about 1,700.

When Hecker finished laying out these principles before an industry group, a listener rose up to ask a real-world, downer-type question from his company's experience with Energy Department laboratories. "We had mutually agreed on a cost-sharing formula," he said, "it was an area within the laboratory's core competence and the laboratory wanted to work with us. But it then said, 'We don't have any money.' Was that because we're a new industry, because we're not a defense company?"

Very difficult, Hecker acknowledged. "The Department of Energy encourages us to do these programs, yet they are expected to be funded in a peripheral way. So it's very difficult to tax existing programs. Just now it is being recognized in the government that we need more than a mandate to do commercial work. Funding is necessary." While Los Alamos, Sandia, and Livermore are best known as nuclear weapons laboratories, about a quarter of their work gets into civilian energy research. USAF's Wright Field has been strictly military. Hence Dick Paul considers the Ohio centers necessary intermediaries in moving into the unfamiliar commercial arena. His Air Force world is long-term, looking for technologies to make weapons, sometimes off-the-wall like his earlier work: running research in command and control for the Strategic Defense Initiative, or Star Wars space missile defense program. But Dick Paul is also an electrical engineer from the University of Missouri and thus can understand the commercial viewpoint: shorter term, profit driven.

Besides the Edison centers network and the Ohio Advanced Technology Center, the state has another participant, the Ohio Technology Transfer Organization. Created about four years before the Edison centers, this organization comprises almost thirty colleges and universities in the state dedicated to working one-on-one with business. Wright Laboratory's Office of Research and Technology Applications responds to requests for help from the university network as it does to the Edison centers.

"Our contribution," adds G. Keith Richey, technical director at the Wright Laboratory, "is normally time and materials. Companies are also contributing time and materials. So no money changes hands. That's why they are called cooperative research agreements." Results, if they turn a profit, are split between parties, but not always 50-50. Money from these comes back to the laboratory, and some can go to the individual, up to $100,000 a year. For a government employee especially, this is not small change.

State research and industrial development centers are hardly the stuff of headlines. Ohio's is more integrated than most and does push technology transfer. Pennsylvania's Ben Franklin partnerships have been singled out, too, as ahead of the pack. These team a company with an academic partner of its choice for either technology transfer or development, and the state looks for payoffs in job creation or in general sharpening of its industrial competitiveness. Professor Nagel of Lehigh, in testimony noted earlier, told a House subcommittee hearing[5] that the Ben Franklin Center at Lehigh had helped company development of more than 400 new products or processes. Pennsylvania's program, he said, "has been replicated and cloned by many other states, some because they saw it in operation and some because they thought of it on their own." Perhaps by no coincidence, both Pennsylvania and Ohio are rust belt states with incentives to revitalize their indigenous industries.

Pennsylvania's Ben Franklin partnerships, as winners for state, industry, and university as Professor Nagel painted them, have missing ingredients. His views are the other side of Frank Moore's skepticism of overemphasis on basic research, that adequate long-term basic research funding is a missing ingredient. So is recognition by state and federal agencies that what's needed is a mix of technology transfer, applied research, and fundamental university research. Hollings centers and industrial consortiums like Sematech are important, but answer only part of the need, and Professor Nagel adds: While Lehigh may be one bright star, there is a lack of funding for centers of excellence, for the Lehighs of the future. "There is a lack of funding for basic creation of organizational entities at lots of other universities, and for strengthening the ones that exist," he continued. "The things that we can do five or ten years from now are not being seeded at the rate that we are going to need them seeded."

Other states have taken approaches, some different, some similar. Washington started a technology center for training advanced students, and for research and development of commercial technology. Michigan established three research centers of excellence, one beamed toward computer-controlled manufacturing, automation, and industrial processes.[6] Networking is starting among states, for George Kozmetsky's Texas center is talking with the Michigan manufacturing centers. Maryland operates regional technology extension centers, adopting an idea from the land-grant college agricultural extension centers, and not long ago created a Maryland Information Technologies Center as a consortium aimed at digital systems. Wisconsin has the industrial GreatLakes Composites Consortium, which was awarded a $50 million Navy cooperative agreement to manage its Center of Excellence for Composites Manufacturing. Perhaps it is no coincidence that Representative Les Aspin, the chairman of the House Armed Services Committee, hails from there. The consortium has three classes of industrial members whose fees range from $6,250 annually for a nonvoting associate membership to $250,000 initiation fees and as much as $1 million over time in project costs and investments.

On the Other Hand . . .

State research centers are not panaceas. North Carolina in the early 1980s started its own state-financed and -run microelectronics research center near Research Triangle Park. That is near the university-industry, high-technology Raleigh–Durham–Chapel Hill complex. Since it was also proximate to the state capital, there was, amid the fanfare of a formal dedication rite, a chorus line of state and local politicians bowing and smiling and waving, making rosy predictions. Five years later the center is hanging by its thumbs, struggling to stay in the state budget. Down the road, the Research Triangle Institute, a pioneering concept in research centers, survives nicely on revenues generated through contract research with the military, the U.S. space agency, and industry. True, it needed state political leadership, especially from then-governor Luther Hodges, a former U.S. Secretary of Commerce. But private capital provided the money. Watching the state microelectronics center's travails, Research Triangle Institute likes its own situation best.

Research Triangle is symbolic of synergism. Its director, Tom Wooten, makes the point that the government-industry-university triangle it represents is far more important than its geographic site. Nothing much happens by itself in technology development or in technology transfer—or competitiveness. Interaction, furthermore, has to work on several levels. Technology transfer, Wooten found out early on, is synergistic, people sensitive, and a pull process, not a push. NASA set out to push when Congress mandated at its inception, as it has with the later statutes affecting the military, that the space agency transfer technology to the outside. "When we first started working with NASA," Wooten recalled, "we worked in the field of medicine. We tried very hard to provide the medical community with information about NASA. We soon learned that we and a lot of other people had a simple-minded notion, that you can take a box with a ribbon around it and say, 'Okay, take this solution and solve your problem.' It doesn't work." The reason is the same one Frank Moore found out about in his Ohio materials center. "Industry works in a problem-oriented way," Wooten said. "If you come in from left field and say, 'I got this great idea,' the chances of finding all the ingredients to make that work are limited, at best." So Ohio and North Carolina are on the same track: start with a practical, pressing problem, then dig up technology with a practical solution. Don't offer technology to industry, no matter how glittering, and expect industry to snatch it up if it doesn't happen to solve a problem. Most research people totally underestimate the industry-laboratory gap, Wooten is convinced.

Intermediaries are indeed the grease in transferring technology from the laboratory power train to the industry driving wheel, something Wooten agrees with, too. Perspective is important as well, for technology transfer is as much a social problem as a technology problem. A government researcher wrapped up in his experiment might be flattered if there is commercial interest in "his" technology, flattered in an academic sense. But he will not bother to push an application outside his mission, especially if government raises arbitrary barriers, as it will do. An engineer in industry has no intellectual interest in the technology or where it came from, no patience with the scientific temperament. Only solving his problem matters. Some agency or some individual has to be a midwife in birthing technology transfer. "So you need a transfer agent," Wooten said, "whose got enough breadth to speak both languages, who knows the problems that have to be solved, who can facilitate. It has to be a pull process."

Industry, with existing facilities and processes tailored to a technology, may not want to hear anything new anyway. Sometimes it takes the new boy on the block. Khanh Dinh, a Vietnamese refugee, tapped NASA's technology bank for a microgroove technique to make the heat pipe at the core of a new kind of air-conditioning system. "NASA has a wealth of such technology it will transfer free of charge," he says.[7] Industry does pick up on ideas from a magazine built around NASA's exploitable technology, but diffusion is slow and subtle.

Knowledge versus product is a thread through all of technology. Often overlooked is the fact that universities, like NASA, may be great generators of innovative ideas, but they don't develop products as a rule. Neither do military laboratories, though they may claim otherwise. Jimmie R. Suttle, a Ph.D. from the Army Research Office at Research Triangle Park, did a tour in the Pentagon before retiring. "I'd go to briefings in the building," he recalled wryly, "and some guy would show up from the Naval Weapons Center and say, 'We developed this or we developed that.' When we would start pinning him down, he would admit it was done under a contract. They would have an in-house scientist working on the project, but I equated it more to having a smart buyer system, not that the laboratory actually developed a product."

What he was getting at is that industry develops products. At North Carolina State University where he is vice chancellor for research, a few miles down the road from Research Triangle Park, there is a new research effort underway that recognizes this proposition. The university, a land-grant aggie kind of tech school, has fallen heir to a thousand acres of open land across from its campus, in spitting distance from Raleigh. There it is building its Centennial Campus, a research park housing both university and industry researchers. Proprietary barriers will be respected, but by and large the goal is to make interaction easy, to encourage industry engineers or scientists to walk across the lawn to talk with counterparts in the university's faculty or research staff. As a sign of the times, the first industry tenant is not an American company, but Swiss-Swedish: ABB Asea Brown Boveri.

Because of this industry orientation, North Carolina State shot up from twenty-third spot at the beginning of the 1980s to, at the end of the decade, fourth in volume of industry contributions to university research. As with military funding, though, the squeeze is on corporate research grants to schools. So those like North Carolina State that court industry programs, rather than regard them with suspicion, are likely to command larger shares of a smaller pie.

From imagination at one end, state research programs, with or without university participation, are perfunctory or imitative at the other. Arizona, for example, approaches research parks more as a way to sell real estate than with any strategy other than that its warm winters and mountain scenery will attract industry. Instead Arizona was an also-ran in competitions for sites for Sematech or the supercollider.

Nevertheless, research and the states have a seed that is going to grow much larger. States are eyeball to eyeball with industry and with the ravishing of local jobs by foreign competition. They will be on the leading edge of the economic impact of military base closings and contractor cutbacks. If federal research declines, states and regions will feel the pain. Should there be a federal industrial or technology strategy? Put that question to Frank Moore, and he answers: "When the Japanese have a meeting, there are a lot of people sitting in the room—academicians, government, and industry. They make a decision that they are going to take the electronic chip market from the United States of America by whatever means they have to use." Subsidy, fine, as long as they wind up with the market. "Somebody in Washington should do the same," Moore said, "so that the United States of America is not only going to go to the moon and build space shuttles, we're going to take the TV market back. They're wrong about not picking winners and losers. You make winners." As the United States did with computers.

Put it to George Kozmetsky and there is a different emphasis. The federal government, he says, is out of touch with what is going on in industry. Particularly so is its lack of a clue to the internationalizing of industry, the new competition that is part networking, part cooperation, and part traditional competitive conflict. Whether the federal government understands commercial technology is questionable. Rather than a twenty-first-century counterpart to a federal defense industrial policy, Kozmetsky foresees state or regional technology venturing, a race among those within the United States and without for the technological, and hence industrial, prize. Federal money may help in some cases, but the national government, if it continues to bumble along worrying about picking winners and losers, will wind up a bystander.

6

How the Rest
of the World Works

Europe and Asia take for granted the kind of government support for industries that Americans often find distasteful. Not only did they support defense during the Cold War, but also civil industrial technologies in the post Cold War. Europeans were good at working out credit for defense orders and at quid pro quos: We'll buy your hams or solve that little diplomatic impasse if you buy our radars. Governments in the rest of the industrial world work at fostering industrial development whereas America's professed hands-off relationship often stops with regulation. During the Cold War, Europe applied the defense approach to civil technology, with government underwriting research and development. Precisely government support will come to dominate global technology as defense shrinks worldwide and nations look elsewhere for ways to drive industrial competitiveness.

Nevertheless American-speak about hands-off and free trade for industrial development or export sales does have a forked-tongue flavor. After all, the Pentagon has a foreign military sales operation that can be most persuasive in pushing its own interests. The United States farms out a negotiated portion of subcontract work to companies in the customer country as an incentive to them to buy American. Beyond that, it opens the American military logistic network to foreign arms buyers, whose orders share in the overhead and reduce prices for everybody. Commercial exports can help, too. Best known is the export financing of the Export-Import Bank, a quasi-government institution established to compete with European government financing of arms to help out sales to countries with less than stellar credit ratings. Defense financing did expedite development of the jet transport and the electronic computer.

Of overriding importance in military export sales is access to leading-edge American technology. Japan has made several coproduction deals, where it built American aircraft like the Air Force F-15 in Japanese plants, learning American aircraft technology in the process. America sold Japan its Delta space launcher technology to give Japan a base for its own family of these vehicles. Technology transfer of such magnitude generated a watershed controversy in industrial and technology circles in the United States and culminated in Japan's plan to build its own F-X advanced fighter aircraft using American F-16 fighter technology. After much debate in Congress the program went ahead—with a critical post–Cold War proviso. Technology transfer was to be a two-way street: U.S. F-16 technology for Japanese manufacturing technology. No question, the old rules are changing.

The World Cuts Back

As American defense spending goes down with a collapsed Soviet Union, so does defense spending in the rest of the world. European companies are going through the same exercise as those in this country: slimming down, looking for commercial applications of defense technology, for Third World military sales. France's Dassault in the last two decades sold its fighter aircraft briskly outside France, including its first-line Mirage F.1 fighter to Saddam Hussein. But Mirage exports dropped steeply after 1987. While trying to expand its business aircraft sales to take up the slack, Dassault has trimmed its employee roster by 20 percent. While layoffs are not exceptional in U.S. industry, they have been rare in Europe, where the unwritten social contract is a job for life.

British Aerospace bought the Rover Group to diversity into the automobile business and raised civilian sales to more than half its total revenue. Rover proceeded to lose money in 1991 while British Aerospace defense profits held up in slow times, perhaps helped by layoffs of 20,000 workers. Like its American aircraft engine counterparts, United Technologies and General Electric, Britain's Rolls-Royce is laying off—6,000 employees in 1991—partly because of declining military sales. Germany's aerospace companies merged into motor car builder Daimler-Benz is aiming at higher civilian sales. Britain's GEC-Marconi is producing TV satellite dishes.[1]

Britain's government is approaching defense spending differently from the United States. At lunch in a club formed to trade on the concentration of defense company offices near Washington's National Airport, Peter Hearne held forth on the coming changes. Peter Hearne grew up in war, as a schoolboy watching Battle of Britain Spitfire and Hurricane pilots scramble off grass fields to intercept German bombers. A veteran of the Cold War, he was running GEC-Marconi's American defense acquisitions. Britain's Treasury dominates defense policy making there, and Hearne's impression is that if a future commitment costs more than a specific threshold, then it's unnecessary, whereas the United States, he believes, will stick to its military commitments, treaties, and alliances around the world in future years, albeit in slimmed-down fashion, Britain—and the rest of Europe—will shrink defense more and turn spending inward.

As for shifting research and development into civilian technologies, like the European Community Eureka program, the financial bind will operate here as well. New taxes are as repugnant in Britain as they are in the United States. So Britain, in Hearne's opinion, will be trying to reduce other government expenses as well as defense to stay with the bounds of tax cuts that came in the 1980s, perhaps, like those in the United States, that were too optimistic in the aftermath of the way governments continue to spend. As in the United States, health care is in trouble, and the government would like to divert defense and technology money into that sector. Britain's options-for-change studies are more or less along the lines of U.S. introspection about industrial competitiveness.

Then there is the Alvey experience, named after John Alvey, a defense scientist, who was the first technical director of the program later named for him. "Alvey," Hearne said, "was started by friend of mine from our company named Laurence Clarke, a research guy very much orientated to new ideas in computers, rather than making money out of existing ideas." Along with Brian Oakley, a GEC colleague, Alvey had the vision of turning around British information technology. At the time, Britain was actually manufacturing only 5 percent of its computer hardware. The rest was imported. "Their idea," Hearne recalled, "was that a whole lot of new and brilliant ideas would be developed, implanted in British industry, and products ahead of the rest of the world would be exported. Sounded great."

Oakley and Clark turned into civil servant pumpkins in the Department of Trade and Industry, setting up research programs in government and industrial research laboratories. As Hearne noted from an industrialist's perspective, "The time scale of the ideas they were financing was so long that we could never really get the pull-through effect into developing products. By the time they got into industry, other companies might have done them independently." As in the United States, industry really wanted to learn how to make a word processor at half the price or get the device to do something quickly, not do research for the sake of research. Alvey money tended to go to chip manufacturers, which didn't help the information technology people particularly.

Defense, which was Peter Hearne's interest, was not neglected—in particular, digital map displays. Hearne was "invited" to share two million of a four million pound program cost in that technology. "Most of my two million pounds," he said, "were to go into helping Plessey or somebody to develop some graphics chips." Ridiculous, he responded. Texas Instruments sells graphic chips. True, less advanced technology might mean buying four chips instead of two. But Hearne's two million pounds would, that way, put a salable product into the market. "I was told this was un-British," he went on. "Getting the technology right was what it was all about, not a product with imported chips." To Hearne, what it was all about was getting products right. Grass is not always greener in government collaborative research and consortiums pastures.

In that, Hearne has a lot of agreement from industry, that development has to be customer- and market-led rather than technology-led. If industry is going to participate in the funding, then, in British idiom, there has to be a quick pull-through; in American idiom, a fast payback. Industry needs new products, sales, cash flow for the next product generation.

"Otherwise," Hearne said, "it's just inviting industry to share in putting money in the infrastructure." To Peter Hearne, Alvey is a good example of what hasn't worked. "Sort of a punish-the-innocent and reward the guilty," he said, a wry reference to the fact that those running it were decorated. Post-Alvey, Britain's information technology industry, was not better off, and supporters argued that it really needed another five years of funding to make things happen. Oakley himself, who also served as an Alvey director, says that, in trying to overtake Japan, the program misunderstood how Japan works. "Their research centers tend to be staffed by people from companies on short-term secondments who then go back to their firms to develop the technology for the market, fighting furiously as they do so. In the UK, staff are allowed to stay, get stale, and grow old." As for Alvey, he added: "The UK was foolish enough to think that prosperity can be based simply on research. . . . But governments must create the right industrial climate."[2]

Where there are Alveys, there are also joint research programs like Esprit or Jessi. Esprit is the European Strategic Program for Research in Information Technology, which followed Alvey in the then-gestating European Community of 1984. Esprit has to be taken seriously, simply because it will potentially have $6 billion invested over 10 years to fend off American and Japanese computer inroads. Both Clarke and Oakley were involved in Esprit, which is controlled by industry and which earned more popularity and better marks in Europe than Alvey did in Britain.[3] Europe is better off competitively than it would have been without Esprit, industry agrees, and the program has stimulated research projects in individual companies that might have been passed over otherwise. Jessi, the Joint European Submicron Silicon Initiative, is related as a research program to develop the electronic chips as the basis for improved computer and communications technology. Its $4 billion multiyear budget is a large figure in government-industry commercial research, but Europeans worry that it is tiny compared to what Japan spends on chip development in a single year and too small to fit Europe's technical ambitions. All these programs underscore the cultural differences between the United States and its international competitors. Faced with a problem, Europe collaborates; the United States competes domestically and expects the winner to carry the technical torch into export markets. Europe does not care for outsiders in consortiums, but there are exceptions. IBM, an American hallmark but with a research base in Europe, has been admitted to the group.

In the smaller picture, Peter Hearne's GEC-Marconi is looking at civil applications, like air traffic control. Paradoxically, one of the potential markets he visualizes for defense contractor skills, though perhaps not contractors themselves, is in rebuilding the fragmented Soviet economy. Former defense research chief William J. Perry has spent the last year or two trying to get U.S. companies interested in converting Soviet defense plants to commercial production, perhaps an easier transition for them because they have traditionally used defense factories for building all kinds of civilian goods. "Lacking a financial infrastructure, looking at Soviet political instabilities," Perry finds, "U.S. companies hesitate." Potential market, sure, a country with 300 million people and no established brand name. "It's the usual balance of greed and fear," Perry observed. "Up to now, fear has been the winner."

A more specific opportunity may exist for certain U.S. defense companies. Compared to U.S. war reserve stocks of ammunition of a million metric tons, says Bill Schneider, who advises companies on foreign trade from his State Department background, the Soviets have seventy million, some dating back to the Russo-Japanese War at the turn of the century. The notion that the Soviets are worse than the United States in squirreling away military equipment is true. "Soviet Republics want to demilitarize a lot of this stockpile," he says, "and, unless you want to screw up the atmosphere, blowing it up is not the way to do it. Doing it in the green manner is a sort of medium-technology disassembly job, denaturing noxious parts, pulling out the electronics. Just anybody doesn't go sniffing around these things that have tens or hundreds of pounds of high explosive." The United States has invested millions of dollars in a demilitarizing plant on Johnston Island in the remote central Pacific, but the Soviets, he says, can't even get a plant built.

Two trips to Eastern Europe and the Soviet Union for Donald J. Atwood, deputy secretary of defense and a long-time General Motors executive before that, brought back skepticism about Eastern Bloc defense plant conversion. A Czech tank plant was building shovels for front-end loaders for a company in Western Europe, business it had won by meeting the Western company's price. "Are you making any money on the contract?" Atwood asked. "Who knows," his guide responded. "We don't really have any cost-accounting system." At a Soviet shipyard in St. Petersburg, the conversion product was a coffee table on rollers. Was that a product of extensive market research and cost analysis? No, said the tour leader. "Someone in Moscow told us that's what we ought to be making." Not many westerners grasp how little the Soviets understand marketing, cost accounting, or distribution, for, as Atwood went on: "All they know is that they have been doing for years what someone told them to." As an avenue for U.S. defense contractors to develop conversion business, the Eastern Bloc is an iffy proposition. Conversion would have ben ordered by the old centralized Soviet bureaucracy; with no familiar directives from Moscow now, defense factories in the new confederation trying to find commercial markets are floundering more than they might have otherwise. East or West, conversion is tough.

Otherwise, to use Bill Schneider's description, world markets for defense are melting down. Whether palliatives like extending Export-Import Bank credit to weapons contractors will help in a declining market is doubtful. Perceptions of military threats are that they are fading and nobody is lining up to buy weapons. Commercial producers think with justification they ought to have prior claim on federal credit sources. Besides, Export-Import Bank lending has been persistently controversial in Congress and in the Treasury. Because the bank's lending is part of the federal budget, the Treasury periodically and consistently tries to have its lending authority emasculated. Arguments against Ex-Im lending are that it puts taxpayer dollars at the service of a few industries, like aircraft and computers, that could develop financing themselves, to the profit of private banking institutions.

Defenders of the Export-Import Bank's export financing argue that, first, the government institution operates with its own money. Interest and principal repaid from past loans goes to financing new lending. Until recently, the bank had operated at a profit, but bad years in the 1980s sent it to the Treasury for new capital, reviving past disputes. Supporters also argue that the United States has to meet what U.S. industry grumbles are easy terms and conditions offered by overseas government financing competitors, which can easily outweigh a better product in winning orders.

Both Boeing and Douglas have railed against European subsidy, and the consortium governments did finally back down. Not quite in the same breath, but McDonnell Douglas soon turned around with a deal to spin off its Douglas jetliner operation and sell 40 percent of the separate entity—with its technology, to the dismay of U.S. politicians—to a Taiwanese government-industry consortium. John McDonnell made it a point when the news was out to drop in on Donald J. Atwood, the number two man in the Pentagon, to get the reaction there. While the Taiwanese alternately warmed and cooled to the plan later, the mere prospect had worried Atwood personally because of threatened loss of business and jobs for American second- and third-tier suppliers. Officially, though, he could only insist on assurances that Douglas would not transfer any technology overseas that the government paid to develop, and Atwood says he got them. So much for the evils of subsidy when capital to survive is the issue.

Industrial policy and subsidy are not dirty words outside of the United States and did clearly figure in the rise of Airbus Industrie. Government indeed plays a prominent role in international competition, but Michael E. Porter, of the Harvard Business School, argues that it is a different one than is commonly supposed. At one extreme, government is relegated to passive participation, seated on the sidelines while market forces play out in international competition, powerless. His studies don't support that view. ". . . Government policy does affect national advantage, both positively and negatively. . . . But governments do not control national competitive advantage; they can only influence it."[4]

The Sound of Music

Expensive equipment, like aircraft and computers, gets lots of government attention, helpful or otherwise, even in the United States. Governments other than the United States delve more into civilian technologies and, within the confines of the GATT international trade agreements, push domestic industry along as best they can, nibbling away at market share here and market share there until the cumulative effect can dry up another U.S. industry.

Japan captured a huge share of the consumer electronics market this way. American manufacturers, in a familiar story, have virtually dropped out of building television sets, although television was an American invention. So was high-fidelity audio, if invention is the right word. American nameplates on home stereo electronics, including those of the pioneers, now usually are attached to black boxes housing circuit boards assembled in Asia. Across the border from Detroit in a suburb of Toronto, another kind of government-industry cooperation is aimed at one of the last U.S. outposts in consumer electronics: loudspeakers.

Japan lags behind there, and high-quality English and German speakers faced pricing resistance as the dollar fell. Helped by more favorable exchange rates and trade relations, Canadian speakers lately have moved into the United States and are aiming for a bigger place in the world marketplace. As long ago as 1973, when Canadian loudspeakers were a cottage industry, Paul D. Barton was the first audio engineer to take the unusual step of using the facilities of his country's National Research Council in his design work. Unlike the U.S. government's plethora of laboratories, Canada, with a tenth of the population, concentrates is government research and development with the thousand-odd Ph.D.'s at the council's facilities near its main campus in Ottawa. Speaker manufacturers, like PSB, for which Paul Barton is vice president for research and design, can rent its anechoic chamber, instrumentation, software, and psychoacoustics database built over 20 years to test ideas. With an eye to the GATT international trade agreements, Canada offers the same rental to manufacturers from other countries, and U.S. companies have done so. Barton spends one week out of every six there in a kind of government-industry relationship strange to Americans south of the border. First generating curves on a graph, he ultimately confronts the acid test, a speaker prototype to find out what it can do in a real room with three dimensions. Cycle time is the benefit, for, Barton said, "once the speaker is physically together, I can have it ready for listening in hours." Within a couple of weeks it can be ready for release to the production line.

Next on the Canadian Research Council's long-range plan agenda for loudspeaker research is a program called Athena. Five competing companies—PSB is one—are signed up in a fairly common practice in Canada of sharing precompetitive research, in this case to develop smart loudspeakers. These would adapt to room environments that are often different enough from the one in the store that the purchaser finds exquisite reproduction there becomes rotten at home.

Canada has a system of Industrial Research Assistance Programs for government-industry cost sharing, and Athena is the kind of work they deal with. To sell its usefulness to the Canadian taxpayer, the Research Council calls these partnerships a way to leverage its own facilities and staff base. Similarly, it is making much of the importance of science and technology in supporting national competitiveness—and the word competitiveness is heard just as often in Canada as in the United States. In fact, the council's long-range plan spanning 1990 to 1995 is titled "The Competitive Edge." Its goal is to double the impact of National Research Council programs by the year 2000 and improve the climate for R&D investment in Canada. Promotion like this is responding to the same forces operating in the United States: deficits and shrinking defense budgets or government research and development funding. "In Canada," Barton said, "we're not immune to that either."

Inscribed on one of the Canadian Research Council's brochures is a quotation from Prime Minister Brian Mulroney a couple of years ago: "Canada's place in the world is not guaranteed. Either we meet the challenge of competitiveness and ride the crest of the technology wave in the 21st century or we exhaust ourselves in the rip-tide of history, fading inexorably backwards." As Barton adds: "If you look around the world it seems that Americans are pretty much alone in their attitudes."

Consortiums are old hat in Canada, because the government won't support parallel research in five different places. Industry is left to do precompetitive research, which the government will back substantially. "All the major Japanese companies do precompetitive research, joint research rather than parallel," Barton said. "Americans don't do that. Apparently it's illegal, at least up to a point." Whether America can play by one set of rules and the rest of the world another will be the industrial policy decision of the 1990s for this country.

Beyond the range of the National Research Council long-term plan is the potential for Canada to develop an integrated audio industry, including electronics as well as speakers. Japan in particular integrates well, but outside of IBM, vertically integrated in most phases of the computer domain, American companies tend to specialize in niches. Audio sounds like a niche, but it is an element of a broader technology: media, based on digital signal processing.

Japan already has digital signal processing technology products on American dealer shelves, black boxes that can manipulate the audio signal to make the listener think he is in a concert hall, or a supper club, or a surround-sound theater. "They are just toys right now," Barton observed. "But when it comes to developing and designing specific chips for specific applications, it will be mind boggling." Military funding pushed along digital technology in computers, and Barton can see, with military shrinkage, talent that's now designing military electronics diverted to digitizing the commercial world that isn't already digitized.

The Digital Wave

Digital broadcasting is not so far off. Farther behind, but coming as well, is digital television. Analogous to what Canada is doing on a smaller scale in government-industry cooperative research, Europe is working in this broader field under the tent of the European Community's Eureka program. As in Canada's audio research consortium, Europe's digital broadcasting effort is precompetitive. Not only are manufacturers like Philips and Thomson participating, but so also are Britain's BBC and France's state radio, collectively under the European Broadcasting Union.[5] Eureka is funding another commercial technology program under the related name Archimedes. Besides speaker manufacturers KEF and the Scandinavian Bang & Olufsen, it includes the Acoustics Laboratory of the Technical University of Denmark, which Paul Barton says is the closest thing in the world to what is going on with industry at Canada's National Research Council.

Archimedes and digital broadcasting are connected, although the former has a far more theoretical ring. Again, it is aimed at psychoacoustical responses, having a test listener suspended in a chair in the center of a large anechoic chamber and surrounded by hidden loudspeakers. Signals to these are tweaked to simulate differing room reflections. To do that, digital signal processing software and hardware had to be developed, experience that will not be lost in future system design.

Digital broadcasting also involves psychoacoustics because of a difficult technical challenge. Satellites used to transmit radio signals are limited in how much data they can handle. For it to be feasible, digital broadcasting data must be compressed, getting rid of thousands of data bits used in a compact disk recording, hoping much of it is redundant. Where a compact disk uses over 700,000 bits per second for each one of the two stereo channels, broadcasting needs little more than a tenth of that. There are several systems available to do such compression, including one called Musicam, which the Canadian Broadcasting Company has tested using earth-based transmitters.

Engineers consider data compression vital to the future of digital technology, not only for broadcasting but also for other kinds of transmission as well. In audio, the unresolved issue is whether data compression tarnishes sound quality, differing enough from a compact disk's to diminish its future in broadcasting. There are scores of devices already operating on the signal in an audio chain, and one more significant manipulation might be an obstacle of large dimension. In the meantime, of significance for American technology, the U.S. National Association of Broadcasters is endorsing the Eureka data compression system. Digital video is coming as well and will be more revolutionary perhaps than in audio.[6]

Along with the narrow, specific technology implications of these government-industry, cooperative, precompetitive research programs, there are broader ones. Europe, for example, is looking ahead to its leaving a hallmark on the world's economy of the 1990s, much as Japan did in the 1980s. Furthermore, Europe is in no mood to be, as it feels on occasion, pushed around by the United States. Despite back-pedaling on Airbus subsidies, Europe will keep its commercial aircraft industry humming. Neither Eureka nor Airbus nor Japan's MITI and their government-industry relationships are a bit unusual in the post–World War II global economy. Only the United States marches to the tune of a different ideological beat.

How the other major powers handle research and development and technology and industrial policies,[7] surprisingly perhaps, reflects the United States in many ways. In Germany, states fund the educational system. They also finance nonprofit research and development institutes, but with a more lavish hand than states in the United States provide, covering about a third of the federal-state total. In the German federal government, the Ministry of Research and Technology coordinates projects and funds most civil research.

Although this ministry has a lot to say about German research and development, the legislature, research institutes, universities, and industry all have an informal voice. Still, research and development is more centralized and more coherent than in the United States, where top priorities are improvement of industrial competitiveness and job creation.

Britain is more like the United States than Germany in backing away from national priorities for research and from central direction. And in the mid-1970s, as in the United States, half its government research and development was for defense. How it administers government research and development filters through a labyrinth again similar to the United States, with both Parliament and its special committees on science and technology and the cabinet directing various departments, the key ones the Ministry of Defense, Department of Industry, and Department of Education and Science. Farther down in the tier come committees, advisory boards, and research councils. Both Germany and Britain have one more common element with the United States: sizable shares of research money are gobbled up by administration or by the government's own research establishments. Britain allocated about double Germany's share of its research and development funding for industrial growth in the mid-1970s, but less than half Germany's share that goes for basic research.

Japan's Ministry of International Trade and Industry has become world famous for its planning and management of research and development in Japan. It is not the sole factor, though, for the Ministry of Finance controls the purse. Contrary to what Americans suspect, the Japanese government share of research and development money for 1985 was 21 percent, the U.S. National Science Foundation reported; the U.S. government share, 48 percent. Larger U.S. defense spending is only one reason. Japan simply has not financed research and development in industry—directly. Technically low-interest loans to Japanese companies are financed by Japanese banks, but whether done hands-off from government is a debatable question. MITI dominates the operating level, and it has functions that are scattered in other governments. Certainly the United States has nothing comparable to the MITI system. These functions include regulation of technology imports and exports, foreign investment, and, by no means the least important, patents.

Thus MITI has authority in a field U.S. industry has come to recognize as increasingly important in a global economy where regulations differ; that is, the protection of intellectual property, especially technology and industrial processes. As a funder of research, though, MITI runs third, behind Japan's Science and Technology Agency, which reports to the Prime Minister's office and runs six research institutes and monitors seven others, and the largest, the Ministry of Education, Science, and Culture, a cabinet department that oversees ninety-five national universities and their research institutes as well as national laboratories and independent institutes.

Where Japan and the United States are similar in research and development is the ratio to gross national product. This runs at 2.7 to 2.89 percent.[8] Their split across basic and applied research are similar and both have roughly the same number of scientists per 10,000 workers. Thus the cold statistics convey an ominous note for U.S. industry. Japanese patents in the U.S. system rose from 5 percent to 19 percent between 1970 and 1986 while the U.S. share declined from 73 percent to 54 percent; America's share of world exports of technology-intensive products declined from 28 percent in 1965 to 24 percent in 1985, but Japan's almost tripled from 7 percent to 19 percent. Obviously, there is something right about the way Japan manages its national research and development even though it has misses as well as hits and is not immune to recession.

Europe 1992 and the activation of the European community is likely to wreak profound change on its management of research and development. Superconsortiums like Eureka—the European Research Coordination Agency—are the precursor. Eureka's genesis goes back to 1985 and a proposal by French President Mitterand, which amounted to an extension and more systematic version of a long collaborative research tradition in Europe. By the middle of 1988, Eureka had nineteen nationals as members, the European Community plus Turkey. The list of projects, involving private companies and research institutes, had swelled to 165 funded at the equivalent of four billion U.S. dollars by 1988 and to almost 300 within another year. Not all countries participate in every project. The list of projects could be a counterpart of the U.S. critical technologies lists: biotechnology, information and communications, robotics, materials, transportation, energy, lasers. Americans should not miss the point that new manufacturing and process technologies are clearly recognized; Germany, for example, was putting over 30 percent of its Eureka funding into these areas before Eureka had been long off the ground. Nor should Americans miss the point that Eureka is dedicated to cooperation in civilian advanced technological cooperation with the goal of enhancing productivity and competitiveness of European industry. Nor should Americans miss a third point: that there is no top-down government definition of research to be pursued. Governments put up the money, but the ideas are allowed to bubble up from the grass roots. Finally, still one more point for Americans to be aware of: Eureka has a 10-year plan and hence a commitment to coherence absent from typically ad hoc U.S. government technology work.

In the vein of all advanced research projects, and not just those sponsored by governments, there is always the risk of hatching a monster boondoggle rather than a breakthrough. Prometheus is the kind of project that could fall either way, for it deals with automobile traffic. While research and development might produce clever ways to reduce traffic jams, solutions to these problems also reside with enlightened government planning and regulation. A daughter project, PRO-ART (PROmetheus ARTificial Intelligence), is aimed, a description says, at the basics of interpretation of multi-sensor input, of understanding human-machine interfaces for drivers. Rotund words those, and typical of research that produces bulging reports and not much that's tangible for the taxpayer. Artificial intelligence is on everybody's critical technologies list for its future promise; up to now a lot of Pentagon money has produced more hype than anything else. Still, the idea of training computers to make certain kinds of human decisions will eventually pay off. So it would be unwise to write off Prometheus, for it does have a practical aspect, using sensors for automatic guidance of automobiles.[9]

Human factors, for all its ambiguity, is important in automated automobiles. Hughes in its HE Microwave research is wondering how to get a radar beam, which only travels in a straight line, to react when the road curves gradually. Or, with radar looking off the side of travel, will it see a precipice or a cliff wall and give the driver a warning of a huge obstruction that the human mind responds to with: "Of course I know it's there, but the road curves, you dumb radar." A lot of human factors work has to be done on whether a radar on a car—a cruise control radar—can be slewed to look at obstacles safely and usefully. Obviously Hughes, as a private company, is spending far less than Eureka on this kind of research. And both have a U.S. competitor in the market already with a backup warning device for motor homes.

At a potentially nearer end of the spectrum is the work Eureka has backed on high-definition television. A consortium of European broadcasters, Vision 1250, has been working with preproduction HDTV cameras and has tested a projector for a screen 100 feet wide. HDTV has moved out of the laboratory under Eureka and into the precommercial stage.[10] European progress got the U.S. Congress excited, but not the Bush White House. Europe got serious in 1986 with Eureka-95 to develop an alternative to Japan's National Broadcasting Company NHK high-definition television system. Japan, of course, was first, with a development program investment by NHK estimated at $500 million before the close of the 1980s and with $400 million more of allegedly private industry money.

More than just the technology itself is involved. Advanced television will have a voracious appetite for dynamic random-access memory chips, a market dominated by Japan, and for other products, such as analog-to-digital converters. So if Japan dominates the high-definition television market, its domination will serve its semiconductor industry as well. Then there are the questions of standards. European countries and the United States now have differing and incompatible television and videocassette recorder standards for broadcasting. Like the original rival Beta and VHS systems for videorecorders, a winning standard, one adopted widely internationally, is itself a step toward dominance of the equipment market. More than home television is at stake as well; applications for advanced videotechnology range from a new breed of computer monitors, to more precise medical instrumentation, to precision military intelligence and simulation displays.

By the time Japan had spent nearly a billion dollars in development, Eureka-95 had spent about $100 million. Led by France, Europe followed with Audiovisual Eureka to encourage creation of European programming for high-definition television and to promote Europe's own broadcasting standard. Europe intends to protect its own television receiver manufacturers from Japanese inroads, helped by Europe's developing its own broadcasting standard, and to go after the U.S. market. With Japanese and European electronics sluggers sharing the market, with no U.S. consumer electronics manufacturer in the top global ten, with only one U.S. television set manufacturer left by the end of the 1980s, it is hard to see how the United States could catch up even with a government research and development push. So much for the dangers of letting key industries slide downhill.

Nevertheless, the Bush Administration report submitted to Congress rejected such pessimism. "First," the report said, "many of the technologies that will be essential to the next generation of TVs are also essential parts of the computer industry. The U.S. has world class strength in computers. U.S. companies are also strong in research and development work in flat panel display technologies. Additionally, much of the foreign-based R&D has been aimed at markets with dramatically different wants and needs." That referred to direct satellite broadcasting to home television. America has an extensive, locally owned system of broadcasters, as well as national networks, a potent political force for Washington to reckon with. In another breath, though, the report added that none of the dozen or so U.S. manufacturers of flat-plate displays has any intention of entering the consumer market "because of the perception that Japanese companies had already tied up the end use market and would not be likely to buy display-related product from U.S. sources. . . ."[11]

Flat-plate screens, the kind that could mount on a wall like a mirror, could make or break high-definition television. Judging from a demonstration on Capitol Hill, high-definition television on today's home set screens is better, but not so much better that a noncommercial or nonscientific user would rush to buy one if he had a decent monitor already. Military applications are likely before consumer, particularly with room-size display panels used in command and control centers. Dr. Craig I. Fields, as deputy director for research and later director of the Defense Advanced Research Projects Agency, put his agency behind HDTV research as part of its interest in displays, stressing the dual-use theme.

He told a House hearing:

If the technology were mass produced, propelled by the success of HDTV, considerably greater economies would be realized by De-

fense. . . . HDTVs will consume large quantities of semiconductors.
. . . A lot of people think about high definition TV as an extremely
attractive box for selling the components and technology inside—the
semiconductor chips, the cathode ray tube or flat panel displays, the
signal processing chips and so on. . . . The country whose companies
supply those semiconductors for HDTV will enjoy a significant economy of scale of production and will have the massive revenues needed
for refining manufacturing and the ever-increasing expense of R&D
required for developing the next generation of semiconductor products. In turn, that will give those companies competitive advantage
in supplying semiconductors for products other than HDTV, [such]
as automobiles and telecommunications, and in fact will give them
leverage in gaining market share in those other downstream industries. Further, to the extent that HDTVs evolve to be computers, as
digital technology inexorably replaces analog technology, an evolution that is inevitable even if its speed is debated, the companies that
produce HDTV will have a competitive edge in capturing market
share in the computer industry.

Congress liked his vision. But Representative George E. Brown,
Jr., reflected the impatience that still pervades Congress today over
administration babble about not picking industrial winners and losers. He stomped on the Commerce Department's inaction. "I read
the preliminary report of your advisory committee," he told Alfred
C. Sikes, assistant secretary for communications and information,
who testified alongside Craig Fields, then at the Defense Advanced
Research Projects Agency. "It's rife with analyses of the opportunities but it doesn't come up with any suggested strategies. We can
tailor bills to suit the policy needs of the administration, but we
can't wait forever." And he added another warning: "If you think
the trade balance is bad now. . ."

Nevertheless, the United States has what amounts to a sleeper
hole card. Japan has been broadcasting HDTV for more than two
years. But the Japanese NHK system is analog. Europe is debating
whether to stick with analog or go digital, and analog has the political backing that always makes wrong technical calls possible. U.S.
manufacturers and universities, with a Zenith Electronics/AT&T
team proposing the most sophisticated, are developing digital systems, though more technically challenging than analog. These—
four of six proposals digital—are courting approval by the U.S.

Federal Communications Commission, which is to settle on a standard by 1993. If a digital standard is selected, if digital technology pans out, U.S. manufacturers could be back in the television manufacturing business.

Besides, HDTV, Congress was nettled over the Federal Aviation Administration's bestowing on Sony a $1 billion contract to supply high-resolution displays for air traffic control modernization. U.S. electronic industry groups erupted. But the sad fact was U.S. industry didn't have the high-resolution color television monitor tubes the system needed to give controllers more information, easier to discern. Themselves sniffing for subsidy, U.S. companies complained. Finally, Sony agreed to build a plant in San Diego and keep the jobs, if not the profit, in the United States.

Congress, which normally listens to constituents, wearies of industry running up to Capitol Hill to ask Congress to subsidize its research, crying they just can't afford the up-front money. Yet industry is also slow to pay back government money if sponsored research does pan out. Former White House Science Adviser William Graham put a different twist on the traffic control radar, claiming that it should have been dual use. Besides, to him it put an embarrassing light on a structural problem in the U.S. government. Commerce was thinking about HDTV, Defense about sophisticated displays for secret applications, and the FAA about high-definition displays for use in its air traffic control. What FAA needed was a high-definition display for a large area with high brightness, which is always the hardest thing to get in flat-panel displays.

"While Defense was planning to put money into it," Graham marveled, "Commerce was wringing its hands because it didn't have that much money for high-definition television. Talk about incoherence! There was not even a mechanism to know what the FAA was thinking about, that brought it up to any higher level visibility before the deed was already done." Years ago Defense, the National Institute of Standards and Technology, the National Science Foundation, NASA, and the like could have said: Let's get together, different needs, true, but similar technology. While serving individual requirements, dual commercial use also could have benefitted. "I'd much rather do that," Graham said, "than end up paying any one company for a special-purpose device."

High-definition television in the final analysis should be a pure digital format, he adds, and simulcast in analog if compatibility with old TV sets is a problem. "We haul data digitally now, all the high volumes are on digital systems. We are world leaders in data compression and in microprocessing and digital transmission." Japanese firms have done very well in, say, digital storage, but so have American. So digital technology is a natural for the United States. "Let's don't screw around with these analog systems that the Japanese and Europeans are proposing," Graham argued.

When Craig Fields was testifying at the House hearings on high-definition television, he was talking about putting about $30 million of defense money into research. Then, after becoming head of the Defense Advanced Research Projects Agency, Fields found himself transferred abruptly to another job, the government's way of firing. Newspapers at the time classified him as a high-definition-television martyr, a sacrifice on the altar of industrial policy he favored and that the Iron Triangle in the Bush White House despised. In fact, his troubles were more political and personal than ideological.

One man's tenure is certainly a small issue. But the Craig Fields case reflected a much larger issue of the divergence between American technical and industrial strategies and those of Japan and Europe, where industrial policy and integrated government-industry research is more accepted than at the White House. Japan and Europe, more so than the United States, describe their technology base as a food chain, with semiconductors feeding semiconductor manufacturing, which feeds computers and other product lines. "Cardinal rules of industrial strategy are to build the food chain and never to break it," an electronics industry workshop report concluded. Government is the technological innovator, underwriting initiatives that it then transfers to industry, and moving beyond today's state-of-the-art to next-generation technologies.[12]

Those in industry who sympathized with the battle Fields fought—which was still being fought two years later—think he also lost his illusions about what government can do. Commenting to a Washington group, he seemed to doubt whether the U.S. government would ever step up to any kind of role in national industrial competitiveness. To pick winners and losers in the first place, there have to be choices among technologies and industries. Losing technical races or losing industries makes for having nothing but losers to choose from. With its Cold War purpose gone, its military alliances emptied of substance, its military research machine with an uncertain future, the United States does face a test of national industrial strategy. Can the U.S. government confine industrial or technology policy to regulation while the rest of the world promotes industry and underwrites civil technology? Not likely.

The New Cold War: Electronics

The old Cold War was fought with cocked bombers and buried missiles armed with enough nuclear weapons to devastate the Eurasian Soviet or North American land masses. The new cold war will be fought with technology, too, but in a research and industrial facedown, not with nuclear standoff or masses of troops. Electronics, though upstaged by missiles, supersonic aircraft, and nuclear warheads, was more decisive in winning the old Cold War—and Desert Storm for that matter—than it gets credit for. In the new cold war, it already is the equivalent of NATO's Central Region as the main battleground and it will be just as decisive.

While it is easy to call this new era a technology or economic war—a new thriller takes that line—war is just the least worst way to describe it. The new cold war is more like a technological Olympics, where the gold and silver medals are economic survival, or a Superbowl, where doing all the right things as a team wins, not just passing or kicking or running. Freed from preoccupation with the Cold War, American technology will blend its military mindset of mowing down the economic competition with business gamesmanship of deal making and outmaneuvering. For there will be networking and cooperation and cross-training along with competition, joint ventures amid classic rivalry. Those who say national industries are passé are right in this context; for the new cold war is shaping up along different lines than national boundaries. Not all American teams will be Americans: regions competing, ignoring national political boundaries; industrial consortiums against industrial consortiums, international, not national; strategies shaped in corporate war rooms where the decisions are made on where to cooperate and where to compete. Some call such internationalization national suicide. Yet the real metamorphosis is the obsolescence of the way international politics has been played in the past. Professor Walt W. Rostow of the University of Texas, taking a long look at competitiveness, calls this era the fourth of the technological revolutions that have shaped the global economy over the last 200 years. Not a bad description at all, for it transcends national boundaries.

Any list of critical technologies shaping the future has an electronic core. Electronics will be critical in the defense industry, in military work, and in adapting to civilian markets. For it is electronics that are most amenable to dual use and most amenable to conversion to commercial technology. While the military can hardly find a new supersonic fighter in a commercial catalogue, it can certainly find semiconductors there. Even bigger systems are not out of the question. The Air Force and the Army in a joint battlefield surveillance system custom designed the electronics but bought Boeing 707 jetliners to house them. While it is better to have a specialized military cargo aircraft, off-the-shelf commercial aircraft can do some jobs and could be modified in a commercial kind of program to do others.

Despite all the rhetoric of the Cold War, no country was interested in testing out a nuclear exchange while living memories of bombed cities remained. But Cold War nuclear standoff cost dearly, too, bankrupting the Soviets and straining the American economy as well. No country will want to pay that price again either while living memories survive. Japan figured this out sooner than this country did. Foreseeing the industrial battleground, it laid out an economic and technical war plan, first in automobiles and steel but especially in electronics. Instead of saber rattling, the new cold war will be fought with polite smiles and diplomatic agreements—and technological counterparts of bullets. What the United States can draw on is its experience in making military technology breakthroughs happen, its winning Cold War strategic management. To do so, it must learn how to manage a much different kind of war in which dual use is a microcosm of the challenge. As Professor George Kozmetsky of the University of Texas argues: No more selling American technology overseas for cash; swap it only for critical technology, with electronics as the industrial core. These are the new tactics.

Bell for the First Round

Japan's first medal in the new cold war came in consumer electronics. Akio Morita tells how it began as just another licensing deal, one that his unknown, shabby little company in Japan had a tough time swinging. The patent license manager at Western Electric was too busy to see the head of his company the first time Tokyo Tsushin Kogyo tried to get a foot in the door. Morita finally signed a patent license agreement a year later, to scoffing from Western Electric that the best to hope for from the new technology was an entry into the rather minuscule hearing-aid market.[1]

Seen through U.S. eyes much later, the story sounds like decision time on the bridge in the terminal hours of the Titanic. The technology was that of the transistor, developed by William Shockley and a team at Bell Laboratories, for which Western Electric acted as patent agent. As part of then-government-regulated AT&T, Bell Laboratories was obliged to sell its technology rather than hoard it for its own use and profit.

At the time the transistor license was sold, Western Electric's appraisal was conventional wisdom. Japan's dominant-even-then Ministry of Trade and Industry balked at approving the initial $25,000 license fee because the transistor did not look that useful. Finally MITI was persuaded, but much development was necessary to take what was then a slender technical reed and turn it into a product. Morita lists a few of the problems: wringing more power from the device by reversing its polarity for a higher-frequency device, finding materials that withstood higher temperatures than the original indium coating in Bell's germanium transistor, and, not the least, designing and building for itself or scrounging out suppliers who would make for an obscure company the small knobs, condensers, and the like to put together the miniature portable radio that launched what later became Sony into the world marketplace.

Not the least of the pain for the United States was that Sony did not beat this country into the transistor radio market. Texas Instruments was first, but did not press its advantage. There's a lesson in the transistor radio, which Morita saw in the beginning as the core of a small radio and which the rest of the world did not—including the civil servants in MITI. Even more so was the case of the ubiquitous Walkman portable tape player that Morita had to force on his company's skeptical engineers. His insistence on following his own belief that intuition and experience were better than the fuzz-ball market-focus groups that shape such decisions in U.S. companies is best said in his own words: "I do not believe that any amount of market research could have told us that the Sony Walkman would be successful, not to say a sensational hit that would spawn many imitators."[2]

Japan and Sony did not always make the right choices. Morita predicted a bloody shakeout in the solid-state calculator market, and he was right. "But," he adds, "I must say that, on reflection, I was probably too hasty in making that decision to get out of calculators. I confess that today I think it showed a lack of technical foresight on my part, just the thing I think we have been good at. Had we stayed with calculators, we might have developed early expertise in digital technology, for use later in personal computers and audio and video applications." In the end, Sony had to buy this technology on the outside, not having it available in-house.

Be that as it may, Sony's dive into the transistor radio market was heart and soul of the Japanese thrust into electronics. As an MIT report noted of nascent, insignificant Japanese electronics companies: "They began with small transistorized portable radios, which initially entailed higher risks and lower margins than vacuum tube sets, and therefore were not pursued aggressively by U.S. firms; these radios opened a new market. The cost advantages of these firms in labor and capital then let them dominate the radio market within a decade." From there, much of the rest of consumer electronics fell into Asian hands.

Unfortunately, not all was due to Japanese genius. Despite lamenting America's lost television manufacturing, the fact is that General Electric, after it bought RCA, itself chose to get out of what was still a profitable business. RCA prices were equal to or better than Sony's and its television set quality was competitive. General Electric just didn't think the profit margins on its American television manufacturing were high enough. So it sold the business to the French Thomson electronics combine (which was a bidder for LTV's Vought aircraft and missile division when it was sold). Asian television manufacturing costs aren't necessarily lower; their manufacturers are willing to subordinate profit margins to volume growth because, in their home markets, big sales equate with high quality.[3]

Learning from Japan

Now U.S. electronics manufacturers are following what is called the Sony model. Trimble Navigation Ltd., a California startup in space-age where-am-I surveying and navigation instrumentation, defines it as go after niche markets, be agile, and be preeminent. By agile is meant moving from concept to market in one year, where bigger U.S. commercial manufacturers are willing to settle for 18 months and military developers are much slower. So Trimble's agility, and its approach to military markets, is particularly pertinent to the way military buying is evolving in an era of difficult-to-define threat and lower budgets.

In military electronics and aviation electronics, commercial or military, American manufacturers are as dominant as the Japanese are in consumer electronics. That military-commercial synergism is vital for success in the electronic cold war. Trimble Navigation, a prime example of marrying military and commercial technology, though its corporate culture is adamantly commercial, was one of two suppliers of Global Positioning System hand-held navigation packs that Army generals put at the top of their list for most valuable lessons of Desert Storm. Founded by a former Hewlett-Packard Corporation staffer, Charles R. Trimble, the company carries over Hewlett-Packard's pragmatic and, for the defense business, a refreshingly independent approach to the military market. While it will sell to the military market—in fact it considers that an important market—Trimble is reluctant to accept any military research and development money.

Trimble's TRIMPACK receiver, literally hand size, was, in a loose sense, NDI in Pentagonese. One more of the ubiquitous federal acronyms, NDI means nondevelopment items. NDI were once called simply off-the-shelf, to distinguish the kind of stuff that could be bought out of a manufacturer's catalogue from customized hardware that the military paid a contractor to develop to its requirements and specifications. Later, when the presidential commission studying military buying scandals recommended more of this kind of buying, it became COTS, commercial off-the-shelf equipment.

Under whatever name, nondevelopment items and dual-use technology developed by the military with commercial applications are related but distinct. Nondevelopment was accurate for Trimble's

TRIMPACK in the sense that the company used its own money to design, test, and qualify the receiver for the military market. But it was designed to a military specification to run in a military competition, not just taken off the shelf and painted olive drab to match a soldier's uniform. Unlike the voluminous and rigid specifications of most military work, however, this was more of a guideline. Ultimately, when Trimble's bid was accepted, its own specification, rooted partly in the commercial world, became the final measuring rod in the contract. In essence, Trimble built receivers to its specifications that were initially tested and accepted by the services. Then subsequent units delivered could not deviate from this specification. As Trimble's vice president for military products, Walter C. Melton, explained: "Develop a hand-held navigation receiver that can be submerged in two meters of water for two hours, that operates from 22 degrees below zero to 150 degrees above? Nobody but the military would buy that." No question about it, circuits evolved from the commercial world. Parts were industrial grade, with temperature specifications of minus 40 degrees to 185 degrees, more than pleasure boaters needed. "We put the unit in a parking lot and drove a truck over it," Melton said, "then picked it up and expected it to operate." So it was hybrid, between dual use and military customized.

Why not take military development money for such work? Commercial electronics companies, in their vernacular, don't want the government in their pants. Military financing gives the government the right to open up the company's books, check on the accuracy of the cost figures used in the negotiations over pricing, and pick up whatever else it can glean. It gives the government a claim to ownership of the system design. True enough, once a company takes military development dollars risk goes down; so does the profit—and paperwork is monumental. Risk goes down, but not out. Military business is binary. Once the product is ready for production, the military may decide to have it built, or it may decide not to. While the contractor may not be out any money if the military decides no—although development contracts often are losers, deliberately so to get the business—the contractor has wasted a lot of engineering time. Commercial producers detest that, for even with a less than successful commercial product, the company probably can sell enough to get back its investment in time and money.

TRIMPACK is important for two reasons. For a shrunken military procurement system, it is one kind of wave of the future, the way free markets can stretch shrinking procurement dollars. On top of that, it avoids expensive military specification proliferation. With only its own specification to deal with, Trimble was able to cut the price of the navigation receivers it built when orders exploded with Desert Shield. Pre-Kuwait, the military order for TRIMPACK, or what the military called SLGR, or Slugger, the Small Lightweight GPS Receiver, was for just over a thousand. When the shooting stopped, Trimble's contract called for over 9,000 receivers. Despite three-shift work and horrendous problems with suppliers straining to meet the burst in orders, Trimble was able to drop the price from around $4,000 a copy, not counting peripheral equipment like antennas, to around $3,000. As the proponents claim, there may indeed be found money for the military in dual use or nondevelopment items.

Some of this free-market optimism began to unravel with a new military competition toward the end of 1991 for an advanced successor to Slugger, called, in context, Plugger—in acronymese, PLGR, for Precise Lightweight GPS Receiver. Industry and government debated whether it should be a true nondevelopment item or the more classical kind of procurement with government development money and stacks of specifications and testing. Some, like Trimble, are calling for the nondevelopment-item approach, in two phases. In the first, the supplier finances development of a receiver and qualifies it with the military as meeting specifications. Then, in step two, sealed bids are submitted for qualified equipment. Low price wins.

The government forecast for Plugger is for a production run of 100,000 receivers over the years 1993 to 1998. At $5,000 to $7,000 a copy, that's $500 million to as much as $700 million worth of business. That could mean annual sales of around $100 million, larger than Trimble's entire 1990 sales volume. That's a big enough figure to attract some big names in defense and commercial electronics: Motorola, Collins, GTE, Magnavox, Texas Instruments. Some of them would obviously like to see it a conventional military request for proposals where they excel and where government pays for the development.

For Plugger to survive as a nondevelopment item, there have to be at least two qualified bidders in the second step, the price competition. "If it's competitive, a real two-step procurement, and there are two qualified bidders, then the low bidder wins and the government has told everybody it will accept those prices," Melton said. "Public law permits the military to accept those prices. But the government has also said, 'Hey, if there's only one qualified supplier, then we're going to have to run an audit, and certify the pricing.'" Trimble would rather have real competition and, with its talent in getting cost out of product, beat the others on price.

While such off-the-shelf equipment offers the military faster response, less complex program management, and probably lower prices, there is opposition to the idea in government as well as in industry. Government managers like to control development, even though it may cost more and slow down the product development cycle. There are specialized military needs, too. When the Air Force bought off-the-shelf IBM personal computers as part of a redesign of the Pentagon's global communications system, they had to be sent to a contractor for Tempest work, to quiet their commercial tendency to radiate signals that enemy snoopers could detect. And logisticians want to control equipment coming into the vast military supply system. They want to keep overhaul depots filled with work and hence off the base closing list. They want to keep their own jobs. Perhaps most important, to do a quick-response, commercial-type program means doing something about the layers of specifications in military contracting that are the program manager's security, whether they are necessary or not. Besides these rather self-serving reasons, there are genuine worries that, if industry is left to finance its own research and development, there won't be much staying power. Contractors will focus on the expense and get out of the game, leaving the Pentagon without enough players for real competition.

Early on, in the summer of 1991, Plugger's program managers were not imposing a vigorous and rigorous military specification burden. Unilateral design changes were acceptable as long as performance was unchanged. "Significant?" Melton said. "You bet. For that allows the manufacturer to transfer in new technology as time goes by." With a five-year warranty on the equipment, a bidder can cover that and, if a unit comes back to the factory broken, replace it with cheaper new technology. "We don't have to buy old-fashioned parts, stockpile them, and put the cost of doing that into the original price," Melton explained. "That's a big change in the way government does business."

By the end of the year, Trimble's nondevelopment-item investment risk abruptly escalated and its bidding logic began to unravel. In the original specification, the unified military program office had called for units to take a reading on the satellites for a position fix, shut down for an hour, turn back on to go through security codes to pick up the GPS satellites again—all in 30 seconds. After a bidder's briefing, the specification was relaxed to 90 seconds. Trimble bet on this and designed its electronic chips accordingly, only to learn in early 1992 that the specification had been relaxed again, to 180 seconds. On the grounds that the change was helping a competitor catch up in price, Trimble protested formally to the program office, lost, and took its case to the congressional General Accounting Office investigators.

Disillusioned, Melton argues that monolithic military program such as this will preserve defense business as usual, not open the door to a new kind of commercial, market-oriented military buying. If the services individually could go out and buy their own commercially developed receiver, there would be a batch to choose from on the market. Instead, he fears, the logistics managers will get the single unit they want, but at a price.

To get involved with nondevelopment-items means a multimillion-dollar investment, company money at risk. Again, in the military market there is always the possibility that the military will change its mind and not buy the Plugger after all. Even if Plugger goes ahead, the Pentagon often overestimates how many it will buy, or the program could be delayed for months or years. Or Congress may not like the idea, especially if there is conflicting lobbying by industry, and may cancel the budget money.

How it all comes out with Plugger will be an indication of how serious the military might be in trying to do something unprecedented: asking suppliers to cut costs rather than just canceling whole programs and doing without to meet budget limitations.

As a fitting footnote, Sony, the company Trimble liked to model, jumped into the consumer positioning-system market late in 1991 with a unit to sell for a couple of hundred dollars less than Trimble's $1,500 entry. Like Motorola with the lightweight cellular telephone, Trimble had a technology lead that it had time to exploit. Now it too must face off the competition from the Western Pacific. It is, by bringing out a commercial version of Slugger costing about $1,000.

Was Defense to Blame?

After near-destruction of its country by the summer of 1945, Japan was sheltered by treaty over the next 50 years. Bootstrapping its available capital, technological ability, and industrial capacity into an economic powerhouse through a government-industrial partnership—an industrial policy so abhorred by some American economists and politicians—Japan learned how to build cheap, quality cars. Licensing American semiconductor technology, sheltering its struggling nascent chip industry, Japan built itself into a force in computers and domination in consumer electronics like television and stereo manufacturing. Not to miss a bet, Japan's aggressive semiconductor producers began to sell to the U.S. defense market.

While all this was going on, the United States was spending much capital as defender of the West against the Soviets. Whether the U.S. had a choice in devoting much of its investment to Western defense commitments is doubtful. At the same time, though, the United States was investing heavily in welfare programs Japan avoided: a social security system, instead of leaving pensions to the savings of individuals; guaranteed home mortgages, so that money went into mass housing instead of savings. If America could have invested the difference between its defense budget spending and Japan's in its industrial base, Daniel Burstein argues, "it is doubtful the United States would suffer a competitiveness problem."[4] But Cold War made the question moot.

Clyde V. Prestowitz, Jr., former counselor for Japanese affairs to the U.S. Secretary of Commerce and later senior associate at the Carnegie Endowment for International Peace in Washington, has written a perceptive, but controversial book among free markets and free traders about how the United States blew its international economic lead to Japan without realizing what was happening, without a shot fired. Only with the October 1987 U.S. stock market crash, in his view, was there any widespread recognition of this somber economic fact. As he wrote:

I resigned from the government in mid-1986. AS I observed developments over the following year, I realized I had witnessed a truly historic turn of events. During the past six years the United States had spent nearly $2 trillion on defense to maintain primacy as the world's military and political leader. It had photographed nearly every inch of the Soviet Union and spent billions of dollars to listen to every turn of the screws of Soviet ships. Yet far from being more secure or more powerful, this country was less so. Despite all the hoopla about standing tall, it was not morning in America. If anything, it was dusk. The United States, which had played the role first of occupier and then of protector and mentor, had traded places with its former protegé—Japan.[5]

While Japan tends to get bashed, it was not the only country getting a free ride at the expense of Uncle Sam. Here (Table 7–1) are some typical Cold War figures from a report of the Organization for Economic Cooperation and Development, which compared

TABLE 7–1

Defense Research and Development
(millions of dollars)

	1975	1978	1979	1980
United States	$9,679	$10,564	$10,400	$10,302
France	$1,178	$ 1,333	$ 1,506	$ 1,640
United Kingdom	$1,397	$ 1,627	$ 1,501	$ 1,555
Germany	$ 571	$ 632	$ 649	$ 582
Japan	$ 57	$ 65	NA	NA
Sweden	$ 183	$ 184	$ 156	$ 133
Canada	$ 59	$ 61	$ 63	NA
	Promotion of Industrial Growth			
Germany	381	382	507	564
France	563	416	437	404
United Kingdom	298	178	114	118
Japan	173	151	NA	NA
United States	65	63	69	74
Canada	135	125	125	NA

Does not include public funds spent for defense, space, transportation, etc., that might have commercial fallout.

All figures in 1975 U.S. dollars and exchange rates.

NA = not available.

government spending for defense research and development for Western Bloc countries with that for promotion of industrial growth.[6]

American defense research and development was ten times that of the next nearest nation, and Japan spent less than Sweden. For industrial growth (admittedly definition dependent), America trailed other industrialized countries, including Germany and Japan. Now Germany and Japan are free to reject American ideas ranging from aid to China and the Soviet Union to cleaning up the environment, to tell the United States to take it or leave it.

Both technical society, in the form of the Institute of Electrical and Electronics Engineers, and trade, in the form of the American Electronics Association, took their worries about electronic warfare—commercial, not military—to the House Science and Technology Committee toward the end of the 1980s. Ronald Rosenzweig, chief executive of an electronics manufacturer but speaking for the trade group, pointed out that with Glasnost, the Soviet Union, recognizing its economic disaster, chose to reduce military tension and reduce the arms buildup. To capitalize on a fantastic window of opportunity, the United States can rebuild its industrial base by redirecting a portion of defense spending to industrial development spending.

"In the past 40 years," Rosenzweig said, "the U.S. has been fighting a war on two fronts, and as Napoleon learned, that's a hard way to win. By fighting and funding a war against Communism and the U.S.S.R. on the defense front and against Japan and Western Europe on the economic front, our great nation has suffered economic decline. Our currency is devalued. We have become a larger debtor nation than Third World countries. We are selling productive assets to foreign countries to gain liquidity. We require foreign investment to finance our debt." Over the past 30 years the U.S. economy has benefitted substantially from the by-products of defense technology, but Japan, with a negligible defense effort, has benefitted even more from integrated government-industrial planning. "The U.S.," he concluded, "should treat economic competitiveness like it treats national security—approximately 1 percent of GNP should be allocated to fund major U.S. programs to improve our economic and technology infrastructure."[7]

These arguments overlook what the industrial competitive position of the United States would be if it, not the Soviets, had lost the war. But the burden of defense was real enough. Now the strategy is to use the talent and experience that went into defense for the new cold war.

Chips on the Table

To pick up the story of U.S. pathfinding in microelectronics, the transistor was not an end in itself. Other kinds of solid-state components were invented, and then printed circuit boards were introduced with the wiring embedded in them. Circuits dwindled in size until transistors and other semiconductors that followed were put together in integrated circuits etched on thin wafers. Defense was hardly a burden in this; rather defense requirements fueled the technological explosion.

In 1975, the United States had almost 60 percent of the world integrated circuit market, in terms of shipments by merchant chip makers. Merchants are those chip manufacturers who sell in the open market, as opposed to companies like IBM and AT&T that make chips primarily for internal use. A decade later, U.S. shipments fell to 45 percent of the world market. Japan, which had only 20 percent of the market in 1975, had drawn even with this country. American semiconductor manufacturers, for their part, had only minimal penetration of the Japanese market in the early 1990s, trade agreement notwithstanding.

While this reversal was going on, another fundamental change had taken place. Defense at one time represented 70 percent of U.S. semiconductor use. By the mid-1980s, defense accounted for 10 percent of semiconductors in terms of sales dollars and only 4 percent in terms of number of chips. Defense research dollars did not bring about the invention of the transistor, but they did play a decisive role in the early growth of semiconductor technology. Furthermore, it was the defense market that first put U.S. chip makers into the big time.

Defense's decline in the chip market had national security repercussions. As commercial market growth outpaced defense, chip makers lost interest in cultivating what became a relatively smaller segment. Pentagon research chiefs became so concerned about their disinterest that they dreamed up mega-research programs like one called VHSIC, for Very-High-Speed Integrated Circuits. Conversely, the merchant chip industry became even more vital to national defense as electronics turned into a technological bonanza. In the jargon of the business, electronics was the most leveraged of the postwar technologies. More could be wrung from electronics for less cost than any other development.

Japan's rise and America's fall in the chip market in terms of sales had an even more ominous implication. If Japan was selling more, it must be doing things right, more so than this country was doing. Therefore, technical leadership was also likely to be moving to the Far East. Indirect measures support this. Japanese scientists and engineers began to dominate technical conferences in the mid-1980s, of particular importance those that were refereed to screen the technical significance of the research work.

Still another ominous implication was the erosion of U.S. semiconductor support manufacturers. That is, if chip manufacturing was shifting to Japan, so would manufacturing of the technically advanced and demanding precision equipment to build chips. Foreign control of chip-building equipment is a clear and present danger to the U.S. semiconductor industry. Chip manufacturers overseas not only tailor manufacturing equipment, but they also can steer it away from U.S. manufacturers.

By the middle of the 1980s, the Pentagon's Defense Science Board named a task force to look into the chip situation, and its findings were not reassuring. Using as a bellwether computer dynamic random access memory chips, DRAMS, the task force found that American manufacturer market share for this most important commodity product had gone from 100 percent in 1970 to less than 10 percent in 1987. Worse, the task force chairman, Norman R. Augustine, said, the Department of Defense, which once bought 70 percent of U.S.-made semiconductors, now bought less than 10 percent.[8]

Although the degree of defense dependence on offshore sources for chips was difficult to determine, the task force thought it was significant. In weapons in the field then, the share was perhaps 3 to 5 percent, but in newer ones, it is probably up to several tens of percent. Most of these chips come from Japan, which is an ally and a minor military power. What's wrong, Augustine was asked, in buying chips from Japan if they are cheaper? Dependence on any foreign source is risky, he answered. Politics and national strategies change. Besides, if a source is an ocean away, an enemy can cut the transportation connection.

Had the design of more advanced logic chips, microprocessors and the like, where Japan is weakest, moved offshore as well? That was a good question to Augustine. DRAMs are built in the largest quantity and push the process state the hardest. More complex chips are rising to the stature of complete subsystems in military equipment, and Augustine added: "It's the next logical step for any country that has the capability to build chips to begin to build those subsystems, be they supercomputers or what have you." Admittedly, then the alarm raised in the Defense Science Board report was as much about what was going to happen as about what had already happened.

In broader terms, the American share of the total semiconductor market dropped from just under 60 percent in 1975 to about 45 percent in 1986. Over the same span of years, Japan's share climbed from 20 percent to slightly top the U.S. share. But most of the U.S. loss was attributable to DRAMs.[9] Market here means merchant semiconductors, not the in-house-designed and -built semiconductors that leaders like IBM supply to themselves. These captive operations have nicked at U.S. merchant semiconductor market share, too, further weakening the position of these manufacturers. Not that in-house design is bad, though its research scope may be narrow. When a research powerhouse like IBM can demonstrate, as it has, the possibility of delving down to the single-atom level to build a semiconductor, the United States has hardly lost the long-term semiconductor war. This bolsters the argument that the United States, with its combination of cutting-edge research and high-cost structure, should stay out of the price-sensitive commodity semiconductor business and capitalize on exotic new technology to make high-margin, weakly competitive niche markets.

Unfortunately for U.S. defense interests, Japanese merchant semiconductor manufacturers have invested more of their sales dollar in plant and equipment—35 percent, or almost double the 20 percent for their U.S. counterparts—and more in applied research and development—13 percent as against 10 percent. Japan's view is long term. Japanese companies have a keen mix of cooperative research and trial by competitive fire. Cost of capital is lower than that for manufacturers in the United States, where the price for past wars, and the cost of defending the free world, of generous social spending habits, of environmental and safety regulation, of a monstrous federal deficit have driven up the cost of doing business.

Japan technologically led in almost every aspect of silicon and nonsilicon semiconductors, including work on gallium arsenide materials, except for microprocessors and custom logic devices and optoelectronics, the task force found. America equaled or led Japan in most chip manufacture processing technologies. But Japan led in packaging and testing. As Norm Augustine noted about ceramic packages for high-temperature applications: "There's only about one place to buy that critical component in the whole world; in fact only about one company in the whole world: Kyocera in Japan."[10]

As often the case in viewing with alarm, the task force recommended the federal government put up some money—$200 million a year from the Pentagon budget—to support a cooperative research venture into semiconductor manufacturing for which industry would put up the capital investment. Practically speaking, if the Pentagon had become such a nonentity in the semiconductor market, the Pentagon would have to invest in manufacturing if it wanted U.S. suppliers on hand to buy from. Nonetheless, such research would have obvious commercial application; hence it was a novel way into dual use and commercially feasible technology.

Banish the though that the Pentagon accepted the idea internally. As in the later case of high-definition television, all the industrial policy arguments that erupted in the Reagan years were hashed and rehashed. Despite the case for defense self-interest, semiconductors were predominantly commercial. "There was adamant opposition to having them funded by defense dollars," recalled Tom Christie, then on the staff of the director of defense research and engineering, and the luckless one bucking the party line got clobbered.

Despite the political ramifications, the government-industry consortium, eventually christened Sematech, did stay in the Pentagon budget. Not only was there Defense Science Board support, but there was also enthusiasm in Congress about using the Pentagon for bolstering U.S. industrial competitiveness. When Craig Fields as the new director of DARPA, the Defense Advanced Projects Research Agency, came along later with his ideas for funding American work on high-definition television, promoting in Congress what the Pentagon hierarchy had turned down, the administration had learned from Sematech. "By the time the administration awoke and said, we have to nip this kind of stuff in the bud, Sematech was already past them." Christie continued. "They got their dander up and drew the line on the high-definition TV."

Getting through the Pentagon roadblocks left Sematech with half the government's $200 million annual share that Augustine's task force recommended originally. Furthermore, DARPA's commitment expires in 1993. But Sematech's original fourteen semiconductor manufacturers invested about $125 million collectively to continue what has become primarily research on advanced processes and tools to make semiconductors. These are the domain of mostly small companies, who cannot afford much R&D on their own. So for that alone, Sematech is advantageous for U.S. competitiveness.[11] Significantly, though, a couple of early supporters have dropped out, and there have been critical appraisals—and horror stories—about the effectiveness of Sematech.

Forced out of DARPA, Fields went on to an earlier counterpart of Sematech, Microelectronics and Computer Technology Corporation. Whereas Sematech is a joint government-industry project, MCC, as it is usually called, is a joint private group whose creation required a clarification of antitrust laws in the National Cooperative Research Act of 1984 to allow corporate research consortiums, which are far more popular outside than inside the United States.

In its formative days, Sematech struggled to figure out which of its two kinds of programs to pursue: equipment improvement programs, EIPs, or joint development programs, JDPs; the latter were the original idea for Sematech, to bring advanced semiconductor technologies out of the laboratory for transfer to member companies. Though some members favored the second option, Sematech has veered toward improving performance, throughput, and reliability of lithography tools, etching tools, and vapor deposition tools, those to build chips as opposed to new chip technologies, as having a more significant impact on the entire industry. Outsiders think that's the right kind of focus for Sematech. However, the jury is still out on its effectiveness.

Perhaps so on Sematech. On consortiums, or alliances, or collaborations in general, Craig Fields disagrees: "That's false. The jury is in." But the verdict is mixed. Consortiums have a poor reputation that Fields concedes is more often deserved than not—but because of execution, not the concept itself.

Getting new consortiums started has become a business for MCC. These make it possible, Fields says, to pursue seminal, high-risk technologies that otherwise just would not happen. A bad technology investment decision by a single corporation can become a good one for a joint venture. Admirers in industry—and Fields has some there, if not in the Pentagon—credit him with turning around an MCC beginning to flounder in unfocused long-term research, the danger of such consortiums. MCC is pooling work in artificial intelligence and optical data storage, which has reduced duplication from company to company, and in foreign technology assessment. Reverse engineering of foreign technology, says Fields, "is utterly proper and above board, and it can be very, very lucrative. A lot can be learned fast," as Japan has found. Collaboration in venture capital investments can be useful, too, for building up small companies to serve as suppliers downstream creating an infrastructure.

Adversaries into Allies

Fighting the new cold war may force Americans into an alliance that industry distrusts: joint government-industry consortiums. One academic group noted: "The United States is not used to designing national strategies for helping its industries catch up with dominant producers in the rest of the world. Sematech is an attempt to do so . . . under the cover of national defense, rather than economic necessity, which handicaps it from the start. What the military often wants in a semiconductor chip (unsurpassed performance under conditions of conflict) is not what civilian industry needs (reliability and low cost)."[12]

Sematech has its own manufacturing research line; its own engineers and those assigned by member companies work on applications and functioning of processes. As in the case of NASA, Sematech and the industry-sponsored Semiconductor Research Corporation from which it sprang are investing as well in university centers of excellence. The University of Arizona has one of these, too, specializing in contamination prevention, a vital factor in getting production yields up and costs down. How industry and university cultures will mix in this kind of application-directed research is yet to be seen. Tenured university professors don't put up with the yoke of assigned tasks or projects. Their product is the result of their experiments: data, interesting maybe, but not very tangible.

To Dr. James A. Hutchby, director of the Center for Semiconductor Research at the Research Triangle Institute in North Carolina, this is a very fundamental problem. "Not just with Sematech," he said, "but with any major project at a university. We at Research Triangle have an industrial approach, a management system and a reward system that attract and focus people, a system where there are lines of responsibility and authority. No such thing exists in the university system." A department chairman may have a pot of money to disburse; if someone on the faculty wants a spectrophotometer for 50 K, then the chairman might get something in return. For the United States to win the industrial battle, there have to be applications-oriented projects and teams, people working together, not do-it-yourselfers. "That," Hutchby said, "is a fundamental difference between us and the Japanese."

Research Triangle's electronic research, housed in a low-key Art Deco building, seems to strike a chord with the Japanese. Otherwise Japan's technology focus is very different—though not omniscient. While Japan's successes have exploded in America's face, Hutchby points out there are more recent cases where the Japanese system fell apart. Japan's supercomputer project, which suffered from the lack of applications software, is one, as is its joint optoelectronics project. Recession came in Japan, and its electronics industry, after a couple of decades of market triumphs, was feeling the bite. Japan's vaunted government-industry complex was having trouble finding new technology to turn into product.

American competitors are not writing off Japan's eventual ability to succeed, nevertheless. "We see the Japanese as formidable competitors," a U.S. counterpart said, "not because they may be winning the peak megaflop war, but because of their determined investment in supercomputer usage." While publicizing their new developments, such as their fifth-generation electronics, they are building a technology infrastructure, creating a strong national user base, and have surpassed the United States in providing supercomputer resources for scientists and engineers. Their own market under control, the Japanese will penetrate both the U.S. and European markets without needing radically new design.[13]

Supercomputer aside, Hutchby did know about one of Japan's electronics successes, the LSI laboratory. In the mid-1970s, using a pioneering consortium, Japan decided to surpass the United States, going from medium scale to match U.S. technology in large-scale integration of semiconductors on a chip. Success in that effort came not from resulting technologies, but from each company's assigning back at home base a picked, counterpart team to the one at the consortium. "We like to think of the Japanese," Hutchby said, "as being really collaborative people. But I'm told that researchers in that LSI consortium were highly suspicious of each other. It took them two or three years before the work really started to come together. Sematech is another example of that. We haven't gotten through the consortium woes yet."

Hutchby spent lots of time with woes of that order in 1990 trying to put together a group to work on a silicon chip fabrication technology. Because of its long-range impact, there was a prospect for getting almost $2 million from the government, via the Commerce Department's National Institute for Standards and Technology, NIST. "So," Hutchby said, "we put together a consortium to install our process on a tool built by Applied Materials, an equipment manufacturer, and working with companies like IBM, Motorola, and National Semiconductor, for validation of our processes." Proprietary secrets, often a barrier, did not hamper access to their people and facilities. "Then," Hutchby said, "we asked for cash: $163,000 over two years. If I had letters of commitment from IBM, Motorola, and Applied, I could have gotten that NIST $1.8 million." He couldn't get the money. The equipment manufacturer didn't like large consortiums and thought it could do the work itself. Five times Hutchby knocked on doors in Texas and Arizona, at Motorola, which funds and is one of the major participants in the Semiconductor Research Corporation. "Our marketing strategy," he said, "is go there, hook them, walk them through our lab, expose them to our people, wow them with technology. They tend to get excited about what we're doing and come on board. Well, I have yet to get the right Motorola people inside this building."

IBM did come on board, but, even with its support, the task has been very difficult in the United States. Not so in Japan. "There six weeks ago," Hutchby said, "I met some people in two different companies and the major national university laboratory. A representative of one of those companies, one of the largest in Japan, has been here already." There was talk not only about assigning researchers at the institute, but also about putting up a laboratory here, one of four planned for the United States. The other company planned to send a researcher for a year's tour and the university was interested in collaboration. "For a whole lot less effort," Hutchby said, "we got a whole lot more to show for it there than here. We've been reading a lot about long range versus short range. It's true."

Resistance comes in companies to opening doors to an outside research institute, a nonprofit organization running on soft money. Not Invented Here, the NIH principle, is powerful, more so in this country than in Japan. Japan does protect its technology, however. "Sometimes," Hutchby said, "there is a feeling we're in competition with a company's own corporate laboratories. It's understandable. You have a guy who has certain jobs his group must do, generally with a budget less than he needs. Here come some yahoos from outside and it becomes a zero-sum game. If they get money, he gets less. He's going to fight it. He's not going to pay a lot of attention to what you're saying you can do technically, other than to trash it wholesale."

Universities are a different matter. Too often, in electronics anyway, the technology arises in the university, but it gets no further than as a publication in a scientific journal. That's the university mission: to publish, to train, to understand, but not to transfer technology into a product. Research Triangle tries to narrow the gap, by dwelling on the industrial applications—for example in its work on low-temperature silicon processing—and trying to attract university talent by raising the inevitable associated scientific issues for faculty researchers and graduate students. This way the university gets publications and training—and money—and the institute gets technology to siphon into its industrial clients.

In the broader sense, the suspicions of academics about industry, of industry about academics, of industry about government, and of government about both create an adversarial relationship that does not help to put world-class products into global markets.

Electronics especially is an example of what can go right and what can go wrong with laissez-faire or incoherent technology and industrial policies. While the United States lost the initial rounds of the consumer electronics cold war, semiconductors packed into computers were another case entirely. Computers, spawned with defense seed money, are, like commercial aircraft, another advantage in the U.S. balance of trade, though a declining one, by 60 percent in the latter part of the 1980s from the $7 billion at the start. Even in the commodity class of personal computers, American microprocessors assembled into American boxes still command a leading world market share. America pioneered the supercomputer. American companies, with Defense Advanced Research Projects Agency money, were first to develop the new supercomputer technology of massively parallel processing.

Not without strain, though. Supercomputer builders like Cray and Control Data began to object, at House technology committee hearings in 1989, to DARPA's espousal of this competing technology from Intel and startups like Thinking Machines Corp. "There has been a great disservice done to the supercomputer industry by DARPA with enormous funding for massive parallel computers and no funding to support true supercomputer advanced technology," an industry vice president complained. "This may be the result of public relations 'gee whiz' claims of massive parallel performance and a relatively ho-hum nature of true supercomputers. Like golf, where one drives for show and putts for dough, in our industry we hear about massive parallel exploits for show and use supercomputers to get the job done."[14] These machines are less stable than those of earlier generations and they do need custom software; but there is a middle ground, moderately parallel machines that can use existing software. Defense needs are at stake, for spy satellite image processing, to justify a government research investment. Yet the flak generated makes good reason for politicians to shy off picking winners and losers in the commercial world. In technology, it can't be avoided. DARPA is involved in another big federal technology program to build a $2 billion national supercomputer network, one that is to double the data-handling capacity of the National Science Foundation's NSFNET and would give small companies and universities access to these powerful systems. Again, DARPA has drawn fire as "the marketing arm" of favored companies in the massively parallel processing game.[15] A network such as this will be the equivalent of a nuclear weapon in the new cold war.

American companies would rather first fight over the merits of a technical approach than cooperate on research for a common answer. In the fashion that Cray later joined the new technology with an agreement to use Sun Microsystems technology for parallel machines, the common good is served eventually. This system brought the United States world leadership industrially until about 1970, and it works pretty well still. Whether pretty well is good enough in the electronics cold war is the issue.

Electronics, deep into commercial use and deep into defense, is the technical bridge into the new industrial cold war. Not only is electronics a lever for the United States to create a dual-use military industrial base, but it also is one for this country to pry out a return on its investment in defense technology. As devastating as Japan's thrust into consumer electronics has been, U.S. industry too often simply ceded the market. American industrial, high-end electronics capability ought not to be written off, especially if its defense technology bank contains hidden gold. Like the Russians in the Cold War, the Japanese are not ten feel tall. Opportunity to turn the tide is coming in the new cold war, just when the timing is right, as Japan, Inc.'s heretofore well-integrated system stumbles under recessionary technological and marketing reverses in a field it had come to dominate.

8

Picking Winners and Losers

If Japan won the technology cold war medal in consumer electronics, America has owned them in aeronautics and space. Yet the complaint is familiar that this country spends too much of its government research money to win them. As defense shrinks, whatever ad hoc industrial and technology policies mature, its effects will intensify the issue of channeling federal support away from allied aeronautics and space and into less beneficiated vital areas: materials, electronics, chemicals, biotechnology.

To examine the pros and cons, go back to the scud and fog that shroud Seattle in the winter. In December 1969 there was even thicker gloom inside one of Boeing's otherwise brightly lit engineering bays in its former military complex on Federal Way. There its government-financed supersonic transport designed for commercial airlines was settling under its own personal political cloud. Coincidentally, not far away in King County, Boeing was betting on a new space complex with the latest in huge vacuum test chambers as a winning entry into the government-funded exploration of space. Other companies like General Electric and McDonnell Douglas, sensing new markets for new technologies, bought those expensive vacuum cells as well.

Neither the supersonic transport nor space exploration have fulfilled all those promises of the 1960s, technically or economically. Denounced as lavish waste of taxpayer dollars better spent on earth-bound social problems, they nevertheless remain as bright lights that attract defense contractors looking for a home for their technical talent and experience as defense shrinks. Most important, they lay out the pitfalls of government funding of big technology and product development.

Where the SST Went Wrong

When the government supersonic transport program was conceived, the rationale was a perennial. Foreign competition was ahead of this country in a critical technology of the future. Britain and France jointly were building their own version, the Concorde, and airlines in those heady days were lining up for orders. Conventional wisdom had it that this sophisticated airplane was too costly for private companies to develop. So the United States jumped in like the two European governments with its own, allegedly better product. By that foggy December, doubt was building that the U.S. airplane was better technically, and it was in trouble environmentally and politically.

Boeing had started out with a swing wing, one that moved forward for a better bite of air at slow speeds and then swept back in flight for less drag at high speeds. A swing wing meant an enormously heavy pivot and surrounding structure along with actuators to change the sweep angle in flight. Soon enough the package became so heavy Boeing had to drop the whole idea, one of the reasons it won the government contract in the first place. Because of the swing wing's technical and financial risk, which ranked in enormity with its weight, the government managers of the program had decided to rely on a low-risk, conventional, existing turbojet engine. In fact, the supersonic transport needed a more advanced engine technically to satisfy airline demands for operating cost and passenger load. In hindsight, the government might have better picked the other semifinalist in its competition, a more conventional Lockheed airframe and its less conventional Pratt & Whitney engine.

Gerry Brines, who in the summer of 1990 was directing Pratt & Whitney's engine participation in revived supersonic transport research sponsored by the National Aeronautics and Space Administration, remembered the initial steps. Pratt & Whitney had proposed an engine analogous to one it built to power the famous Blackbird, the just-retired SR-71 reconnaissance aircraft. General Electric, which did not have the SR-71 experience, won the contract. "We had a big meeting here," Brines said, "and a lot of people asked the engineering vice president at the time: 'Why did Pratt & Whitney lose to GE that had basically no supersonic experience?

We were flying Mach 3.5 every day?'" Came the rueful answer: Pratt & Whitney knew too much and it had painted the challenge tougher and more expensive. In fact, it was.

Not only did politicians anticipate the technical problems, but there were also other objections. When Richard Nixon took office in 1969, his Ad Hoc Review Committee divided seven to four in recommending prototype work be stopped pending further research.[1] Other scientists and economists lined up against the airplane as well. More devastating were the outcries from environmentalists that the sonic boom generated by the airplane in flight and engine noise at airports would be psychologically intolerable and physically damaging.

Those beleaguered final days still haunt John Swihart, now retired as a Boeing vice president, but then company chief engineer for the supersonic transport. "Oh, that was just terrible. We were fighting every inch of the way in the Senate. Reaction to the Vietnam War spilling over on us. The kids were antitechnology. Marching into Senator Pastore's office with a television camera, they would say, 'Here's 8,000 Rhode Island signatures. Are you going to vote for that airplane?'"

Program managers are all too familiar with how politics and technology interact. Swihart and the program manager went to see Senator John Sherman Cooper of Kentucky. Like Pastore of Rhode Island, Cooper was a neutral, with no contracts in his state for self-interest. "We had the normal 15 minutes," Swihart recalled, "so I gave him the quick pitch. Then he started asking us questions. We stayed there for 45 minutes. 'Young man,' Cooper said finally, 'that's a fabulous story. The United States needs an SST. Terrific!'" Swihart was pleased until the senator added: ". . . But there's no way I can vote for the program. Your Senator Magnusson put the health hazard sign on the side of the cigarette package, and I'm from a tobacco-producing state." A year later, in March 1971, the supersonic program was canceled.

At the core of the supersonic transport's appeal to airlines—and at least initially it had a great deal—was productivity that comes with passenger load and speed. Of the two, speed historically has been vital in airline growth. If a faster airplane, like the first jet transports, can fly a route in half the time, that doubles its productivity. Speed and newness multiply customer appeal. But the technology has to be in hand. John Swihart himself agrees in hindsight that the U.S. supersonic transport had well-nigh-insurmountable problems in weight, power plants, range, and limited passenger load—all factors that stunted orders for the earlier Concorde. Engineers who felt betrayed by cancellation now agree it would have been a disaster to go ahead with the airplane. Like the Concorde, it would have needed premium fares, and airline deregulation killed premium fares. "I've said this a jillion times in the last few years," Swihart insists, "government has no business in such a program, once the underlying technologies are understood and matured. It's up to the industry to make an economic product." Instead government was dictating airline delivery sequences, trying to "mastermind the program sixteen ways from Sunday."

Keeping the Technology Alive

With the cancellation of the supersonic transport, the government went from one extreme to the other. From a full-bore aeronautical product development, the federal apparatus by the end of the 1970s was ready to drop aeronautics research entirely, at least as it was carried out by the National Aeronautics and Space Administration. Ironically, the space agency had been built on the foundations of an earlier government aeronautical research organization, the National Advisory Committee for Aeronautics. To the dismay of the aeronautics fraternity, space and the Apollo lunar landing had almost crowded out aeronautics by the late 1960s. Yet supersonic transport did not disappear entirely. Commercial supersonics, as the Concorde demonstrated, were certainly technically, if not economically, feasible. An advanced supersonic transport that could fly at standard fares could be a world leader in the commercial airline market.

The concept for a high-speed airliner began to revive early in the Reagan Administration. Aeronautics began to look like an obsolete industry to government budgeteers in the waning days of the Carter administration. "We saw the Carter administration actually determine that aeronautics was a sunset industry," Boeing's Louise Montle says. "They were absolutely going to shut down aeronautical research." She was working at the time for John E. Steiner, who made his reputation as project manager for the successful 727 airliner. Later he had overseen all new airplane development at Boeing. When NASA aeronautics dismemberment seemed imminent, Jack Steiner had become chairman of a policy review committee, working with then Reagan science adviser George A. Keyworth II, that produced in 1985 the first of the "bird books," nicknamed from the eagle on the cover and the subject, aviation. "We got this whole study started," Louise Montle recalls, "to ask about the first A in NASA, aeronautical technology. That A is not at sunset, nor is the industry [at] sunset." Competition was coming, and despite its aversion to industrial policy, the Reagan administration instead devised one, narrowly restricted to aeronautics and space. The first bird book, actually more of a slim brochure, recommended that the United States pursue three aeronautical research paths.

First was bread-and-butter commercial subsonic airliner research as the real core: greater fuel economy and more efficient engines, composite materials and the like—not neglecting improved air traffic control. Alarmed at technical challenges from Europe and Japan to U.S. aviation manufacturing dominance from overseas, the report noted: "This trend, if allowed to continue, will evolve into major national weaknesses substantially increasing the vulnerability of America's position in world affairs." To generate excitement came the second and third paths: a high-speed, long-range transport and transatmospherics, the latter obliquely couched but hinting at flying passengers on the edge of or into space.

"Keyworth wanted something visionary," Louise Montle recalled, "and it appeared that space was getting all the press." Subsonic airliner technologies were essential but lacked showmanship, and even industry was saying there was nothing much new there. "The administration heard that," she said. "Why wouldn't it shut down NASA aeronautics? So I embroidered these three, subsonic research, supersonics, and transatmospherics. Somehow we had to tie ourselves to space."

Behind transatmospherics was the idea that there has to be a better way to fly into space than at the end of a Roman candle. "So I called it a C-TAV," she went on, "civil transatmospheric vehicle. TAV for the military, C-TAV for civil. Jack went off to a policy meeting at Keyworth's Office of Science and Technology Policy with a little sketch in his pocket. They loved it." Not that the concept was exclusive. Britain had been working on a similar idea in its civil space program and the Air Force aeronautics researchers at Wright Field had a spaceplane program. Transatmospherics came out of the black then, to go into the Keyworth bird book and let the Air Force start talking openly about a future space fighter. "Then the next thing I knew," Louise exclaimed, "we had this damned Orient Express."

Orient Express became one of those grabber labels that program ideas need to get into presidential messages. Before long, it, like government research programs often do, took on a life of its own. Orient Express was as an X-designated demonstrator intended to carry on the lineage of the X-1 that broke the sound barrier after World War II and the X-15 that ventured first into the edge of space. At the same time, it became an untraditional, expensive technology demonstrator, a multi-billion-dollar financial burden, shared by NASA and the Air Force, neither certain it wanted the project and one Congress sporadically tried to kill.

Between the first bird book and the second, issued by Keyworth's successor, William R. Graham, in 1987, the National Aerospace Plane became the official name for what started as TAV and Orient Express. For a while it seemed as if NASA had not read the rest of the policy. The message was clear enough that there were two other elements, and with a new national policy, supersonic transport research became more respectable, though not under that name.

Rather it became the NASA High Speed Civil Transport or High Speed Transport program, depending on which kinds of research were carried out, for both technical and political reasons. In the beginning the name was artfully constructed to cover both spaceplane and supersonic transport. Howard Wesoky, who monitored this program for NASA headquarters along the mall in Washington, agreed that there was some confusion of spaceplane with the supersonic transport. "Unfortunately," he said, "when the program was first announced, President Reagan used the term Orient Express. That has stuck with us. We, NASA and industry, don't really see the application of that technology to commercial transports in our lifetimes."

Compared with spaceplane, the money coming along now for supersonic transport research is not awesome. In absolute terms, it is not small: perhaps $300 million in environmental work through

1996 and perhaps $800 million in the second phase to focus on technology itself. Surprisingly, after the debacle of the first U.S. supersonic transport, Congress was receptive. For one reason, U.S. industry's slipping competitive stature is clear on Capitol Hill; another is the smarter way the government has laid out the program this time.

Developing a Strategy

Malcolm I. K. MacKinnon, chief engineer for Boeing's resurrected supersonic transport project team, puts all this in perspective. He does believe there is a future for the supersonic transport as a commercial development. He endorses the NASA strategy: Answer the environmental questions first, questions industry cannot answer itself, before designers like Calum MacKinnon are locked in.

"To see if there was any synergy between spaceplane and commercial transport," he said, "NASA started asking commercial transport manufacturers to do some studies. Slowly, as we did more and more work, we came to the conclusion there was practically none." Synergy looked more promising early on to the other study contractor, Douglas Aircraft, but very high speeds, not even as high as those of the Orient Express, mean special fuels that airlines balk at. They mean literally a hot airplane, one that burns airline paint and emblems off the skin, one that may be too hot to handle on the ground after landing.

Still Donald A. Graf, a veteran of TWA who heads the NASA-funded research program at Douglas, can't stifle his airlineman's enthusiasm for what a successful supersonic transport could do to an airline's competitor. Nor can Malcolm MacKinnon. New York to Tokyo now takes about 14 long hours. A supersonic transport could do that in seven, or perhaps less, about the length of an Atlantic crossing these days. Business travelers yearn for such time saving; airlines for productivity. Behind the new NASA research then is a quest for a much larger airplane than the 100-passenger Concorde with much longer range.

Productivity and affordability will eventually make or break the airplane. Malcolm MacKinnon says that to be salable to airlines, a supersonic transport has to make a profit at 1990's 747 fare levels, which are about 9 cents a mile. While business travelers might pay extra for a faster trip to Tokyo or Tel Aviv, not enough would do so for the airlines to buy big fleets. A market of a couple of hundred airplanes wouldn't be worth the investment on the part of a manufacturer, which could run to $8 billion or even $12 billion.

Affordability notwithstanding, economics are not the most immediate hurdle in NASA's research. Environmental protests, as much as any one thing, defeated the first U.S. supersonic transport and constricted the sale of the Concorde to the national airlines of the two countries that gave it birth, airlines that had to buy it. Engineers were locked in, too late to do much to meet the environmental issues, real or specious, when they developed.

While least defined, stratospheric engine emissions from the supersonic transport are potentially the riskiest because of what they might do to the earth's protective ozone layer. Charlie Jackson, later head of the division at NASA's Langley Research Center in Virginia whose arsenal of advanced wind tunnels house a good deal of the supersonic transport research, lived through the first U.S supersonic transport program at NASA. "I was a working engineer on it," he recalled. "But the airplane didn't get canceled because it was a dog. It got canceled for environmental reasons."

"Absent environmental criteria," Jackson said, "then we have an emotional issue. There's no way to deal with that." Ozone seem the most intractable now, not so much for technical reasons but for the lack of benchmarks. Jackson's deputy, Allen Whitehead, manager of the high-speed research program at NASA Langley, elaborated: "Is any ozone destruction palatable? Or are we looking at half a percent a year?" NASA empaneled a special working group of university scientists to develop a working three-dimensional model of how ozone lives and dies in the upper atmosphere as part of its generic, not product development, research.

Engineers want to establish some milestones on the progress of the scientific working group. "We're trying to get them diverted from total focus on CFSs to look at this problem," Whitehead said, "because it's a limited community out there. Maybe a dozen models, most two-dimensional. One of the first questions I asked when I got this community together was, 'When can you guys give us a feeling for the uncertainty of your models?' There was this blank response: 'We don't do that.'" So came the traditional scientist-engineer dichotomy. Yet the scientists will have a big role in determining the viability of the airplane.

Once designers have an ozone target to work toward, then the scene shifts to engines where the oxides of nitrogen are generated. Pratt & Whitney and General Electric are working under NASA contracts to test ideas for designing burner cans, the section of a jet engine where combustion takes place and air is heated and expanded to spin the turbines that make the engine work. These ideas are loosely analogous to what carmakers have done with engine fuel-air mixtures for better mileage and lower emissions.

At NASA's Lewis engine research center, which is sponsoring engine emissions research, Robert J. Shaw warned that they were not the only barrier to an acceptable supersonic transport. Another major problem is noise at the airport. Joe Shaw was a young engineer at Lewis during the first attempt at a U.S. supersonic transport, and he remembers that engine noise was a tough problem then. "At times," he said, "I think we talk so much about the emissions problem that we forget that the noise problem is damn tough, too. A lot of talented people worked for many years not only in the first supersonic transport program, but also in later research, and didn't solve it."

To deal with the third environmental barrier, sonic boom, NASA Langley is trying psychoaccoustics. Although there are ways to design a low-boom supersonic transport, the airplane gets too long to build and fly. So in a small concrete blockhouse in a corner of the accoustics laboratory, NASA built Fort Boom. In the walls are an array of 15-inch woofers, like those that make the deep base notes in loudspeakers sold at the local stereo store. Dozens of average citizens, like those who might be hit with a supersonic transport's boom, are subjected to pairs of low-fi, low-frequency thumps, like the sound of a bored small boy kicking his feet on his bedroom floor overhead. From test reactions, NASA will try to find out whether any level of sonic boom is tolerable and, if so, how loud. An overpressure of two pounds per square inch, like the slamming of a door, might be acceptable to the public, designers hope. Otherwise the airplane would be restricted from flying over land, killing its speed advantage and customer appeal.

Why of all things should the U.S. government now invest hundreds of millions of dollars to get ready for what might be an illu-

sory age of supersonic transport? Isn't this picking winners and losers? Isn't this the worst kind of industrial policy? Yes, but . . . In
its first study, Douglas estimated that capturing the market for an
environmentally and economically successful supersonic transport
could mean for the U.S. economy a $100 billion swing in balance
of trade, 200,000 jobs, and $500 billion to gross national product.
With defense contractor layoffs rising, new jobs and markets, along
with fostering aircraft as America's leading export surplus, make
political points.

Besides, there are generic as well as product-specific technologies.
Clever engines are one, adding to American technological competitiveness in its leading export market, commercial transports. To
build one of the novel engine combustors under study to control
emissions and economics, Pratt & Whitney's Gerry Brines points
out, will probably mean an advanced ceramic matrix material that
will have tremendously high development costs. With all the other
engine and aircraft research to do, Brines wondered whether even
the millions NASA was proposing would be enough for the task.
But early in 1992, NASA did award contracts for developing these
ceramic materials. Next, try to get a chemical company to invest
in development of these exotic materials. The earliest anyone can
visualize for launching a supersonic transport is 1998, and sometime in the next century is more likely. To tell a commercial company, with no special stake in aeronautics, there may not be any
payback for 15 years, is a bit much. "They can list a whole lot of
other things they would rather do," Brines said.

Composite material structures, not metal but plastic reinforced
with carbon fibers, will be essential to make such an exotic airplane
light enough to fly at Boeing 747 fares. Military aircraft were the
vanguard in putting an all-composite structure on the production
line—at a premium price. Since then, commercial manufacturers
have been inserting composites into their airplanes in a small way.
These materials from military technology have found their way into
premium-price tennis rackets and golf club shafts and will someday
be a candidate to replace steel in automobiles. Composites, dual-
use technology, are on everybody's critical technologies list.

Given solutions to what are really staggering technical demands,
will airlines buy an airplane that might cost $200 million or $400
million a copy? Given enough productivity and market appeal, they

will. Can any manufacturer afford the development and production startup costs without government help? Taking the most important lesson from the first U.S. supersonic transport, if private capital won't step up to a product's development, its market is dubious.

NASA's supersonic strategy for the government to pay for the most difficult, high-risk, generic research. Then NASA can bow out. Boeing's Malcolm MacKinnon put the correlation neatly: "This has to be a 100 percent commercial program. There is a role for NASA like there always is, to develop emerging critical technologies. They're doing that now, bless their hearts." Without government to sponsor high-risk research, a supersonic transport is not possible. "If we were to build this airplane today," he said, "it would be a dog."

Government investment in long-term research and development, especially if it is in commercially potent technologies with broader applications than aeronautics, is a clear advantage. If the supersonic transport turns out to be unfeasible, it becomes a technology "pork barrel." The California economists who did the study under that name found a disturbing pattern of failure in five massive government research programs in commercial technologies. One, an Energy Department program to develop photovoltaic cells for generation of solar power, was a qualified success. Failure comes with the temptation to move out of generic research and into a more narrow, specific project, which tends to become success oriented. Program people become dependent on and emotionally attached to their research and lose their objectivity about ever-present technical obstacles.[2]

To bypass such weaknesses of the first supersonic transport program, NASA's new research is generic, to avoid narrow concentration on product. Saving NASA's aeronautics research complex is another generic plus. The United States has lost its global dominance in too many other industries to let aeronautics slip away. Still, the overall question remains: Should the U.S. government be investing in aeronautics either to the exclusion of equally important technologies or, in effect, placing a lower priority on other technologies? Will a supersonic transport do enough for broader American industrial technology or competitiveness to justify the NASA investment? One way to look at an answer is that aeronautic research funding to support market dominance in a sunrise industry is not too high. Rather, government effort in fostering industrial competitiveness in general is too low. So the answer is to raise the level of the rest of the research, not squeeze down aeronautics.

A Future in Orbit

U.S. government investment in space is counterpoint to whether the aeronautical sciences are being overdone. Unlike aeronautics, space has a natural national mission: exploring the universe, a modern Columbus voyage syndrome, finding new lands, colonization. Beyond those, it might be a good way for pragmatic politicians to preserve military technical teams, create jobs, keep defense factories operating, and, in the process, plant a bigger foot on technology's highest ground.

George Bush in fact made a feint in this direction. His Space Exploration Initiative goal to put a colony on the moon or send a manned spacecraft to Mars by the year 2025—at a guesstimate cost of $500 billion—is technology policy, even industrial policy. Timing is certainly fortuitous. Military budgets are shrinking, freeing up defense demands for engineers and scientists and for funds. Grafted onto a near-term U.S. space station plan, which includes large investments by Europe, Japan, and Canada, there is a foundation for a long-term commitment of U.S. technical and cutting-edge industrial talent.

Before anyone need worry about the mammoth challenges to executing a manned mission to Mars comes the fact that the administration committed more words than money to this goal. The president's space goals had the same wistful ring as his other slogans: a gentler nation, the education president and a thousand points of light, do-good for the environment, let the unfettered capability of American industry take on the rest of the world. Mars missions of lunar colonies are useful politically because they are such long-term enterprises at the margins of technical feasibility. They are so far in the future and so many technical solutions are needed to bring them off that they can remain fine-sounding long-range policy. With the fiscal 1993 budget, the president implicitly recognized his predicament in space and backed off. Instead he asked for $100 million starting money for more realistic robotic exploration of the moon, and maybe Mars later, before committing to difficult and costly manned missions.

While the United States needs a permanent space program, given its investment there already, a wavering long-term goal of a landing on Mars is not the kind of bold decision that the Apollo moon landing program was, a national goal that stimulated an entire generation of students, engineers, and scientists. Neither is the space station as a centerpiece if its payback to broader U.S. technical and industrial interests, not to mention space exploration itself, is not better articulated in the context of the new cold war technology competition.

"NASA talks in terms of spinoffs," argues Gene Levy at the University of Arizona. "That's not where the benefit is. Apollo's benefit is that there was a focus." When the Vietnam-scarred 1960s ended, Gene Levy was in graduate school. Now as Professor Eugene H. Levy, he heads both the Planetary Sciences Department and the University's expanding Lunar and Planetary Laboratory. The focus of the Apollo decade was on technical creativity, he believes, and a lot of young people were swept into science and technology in its wake. There was a projection from the national leadership and programs that backed it up. In the decades after Apollo, the Arizona laboratory was one of several university laboratories that became quasi-industrial key players in designing experiments for unmanned space missions to both unexplored ends of the solar system; from Mercury closest to the sun, to Venus, a landing on Mars, all Earth-like planets; and to the distinctly non-Earth-like gaseous outer planets: Jupiter, Saturn and Uranus with weird moons of their own. With the president's 1993 budget both praising and backing off from these kinds of planetary space missions, Gene Levy was watching values change to the acquisition of things already created and owned by others, not knowledge. "Whether it was leveraged buyouts or buying oil fields instead of exploring for new ones," he said, "I think that's the loss."

Besides less emphasis on innovation, there was another down side to the end of Apollo, one that hit the universities first. Committee reports in the 1950s and 1960s warned America was backsliding in the training of engineers and scientists the United States needed to compete internationally. Whereupon, with space and defense cuts in the 1970s, jobs disappeared for those who chose technology. Universities couldn't recommend by then that students opt for careers in space science, and they stayed away in droves.

Letters in the scientific journal *Physics Today,* in the early 1990s, were an echo, complaining of the diminished job market for those who were sold on the country's need for scientists in that field. Yet blue ribbon scientific and technical committees and those worried about industrial competitiveness continue to lament the lack of interest in this country in technical careers. How the country seizes the imagination of young people profoundly enough for them to opt for science or engineering depends, to Gene Levy, on the vision projected by national leadership. "One of the most important things the space program does," he said, "if it is done well, as both Kennedy and Johnson did, is to project a sense of priority." By the 1980s, students wanted to go to business or law school. Apollo's era had created a cadre of scientists and engineers. As Levy said: "We're still living, too much, off the fruits of that cohort."

Travail in Orbit

While space still is a wave of the future, can it be a cornerstone of a national technoindustrial policy? Certainly it can create technical stimulus, as Apollo showed. Since then, however, NASA's parochial institutional interests have gradually come to dominate the direction of the civil space program, and not for the better. As a result, NASA is in deep trouble with a once-admiring public. And public policy needs support to succeed.

President Nixon's cancellation of the last three moon landings shook the space infrastructure that grew up with Apollo. NASA still had enough public interest and policy clout to sell the Nixon White House on the idea of building a reusable space launcher, the space shuttle. It also convinced the administration to order the military to stop building throwaway space launchers and use the NASA reusable launcher instead. Then the shuttle Challenger blew up, the shuttle was grounded, and the military went back into the space launch vehicle business with its own boosters, though its biggest payloads crowded out commercial payloads in the limited shuttle missions in the five years after the Challenger disaster while a replacement was built.

NASA had a pretty good budget argument that the United States ought to concentrate its space money on one reusable launcher; never mind that it also concentrated on NASA as an instrument of space policy, not the military. How much risk was entailed in concentrating as well all the country's space launch capability in a small fleet only became obvious a decade later. In the meantime, the shuttle program struggled along underfunded, creating another problem. It soaked up money for space science, though there were striking exceptions, such as the Mars Viking lander. Still it was enough to frustrate NASA's important science and university constituency, like James Van Allen, the University of Iowa physicist who discovered the earth's radiation belts that bear his name, and Gerry Wasserburg, the California Institute of Technology pioneer in radiometric dating of rocks that told us how old the moon was.

Political indifference was partly the reason. At least of equal importance was the evolution of NASA into an organization built around its own institutional needs. In other words, it had become an agency to build the space shuttle, preoccupied with the space shuttle, funding mostly the space shuttle—all the while losing ground with what it was created to do: explore space and make it useful. NASA became so preoccupied with protecting the shuttle that it also began to cut corners on flight procedures.

Compared to the flight test program for a conventional airplane, the shuttle's was cursory: half a dozen glides from a rack on top of a NASA 747 to landings on a broad, dry lakebed in California and four or five launches into orbit. Then NASA declared it operational when it was still very much in a research and development, experimental stage. NASA knew this, of course, and it could be excused for using political reasons to save its only entry for future space races.

As the Challenger accident investigation commission report recounted in obvious dismay, there was intense preoccupation in NASA with getting the first shuttle mission off the ground and back.[3] When the shuttle was then declared "operational," it wasn't really. Loose ends littered the scene: the system was not ready for the routine and scheduled maintenance of a burgeoning flight schedule; spare parts were not in stock; software for a limited-flight development program was not necessarily up to the demands of the two flights a month NASA hoped to launch. Crew training was another potential problem. Astronaut Henry Hartsfield told commission investigators: "Had we not had the accident, we were going to be up against a wall. . . . For the first time somebody was going to have to stand up and say we have got to slip the launch because we are not going to have the crew trained."

NASA had promised the nation that with the shuttle it would have an airline into space. While the commission report gave NASA credit for an extraordinary effort to fly the schedule it did, and especially to retrieve malfunctioning satellites for repair as it had promised the shuttle could do, NASA either did not know how or was not ready to run an airline. Nor was the technology there. Pressures in the shuttle program built up like steam in a boiler. Safety criteria were not met; they were waived or relaxed until the hardware could meet them. Eventually the pressure to meet an "airline" launch schedule, an "operational" launch schedule, meant that engineering warnings were submerged to what euphemistically became management decisions. For at least a year NASA and the contractor had known the solid motor seals that failed in Challenger showed signs of leakage, especially if it was cold at launch time. Yet the issue was closed without fixing the design, with the contractor left to do something more or less on its own. Of course, it was not much of an airline if it couldn't fly if the temperature was freezing, as it was the night before Challenger's last launch.

As a later committee observed, important as manned flight and the shuttle may be, "it was inappropriate in the case of Challenger to risk the lives of seven astronauts and one-fourth of NASA's launch assets [its four orbiter fleet] to place in orbit a communications satellite."[4] A less expensive unmanned booster could have done that job—but without the media attention in having on board the first school teacher to fly in space.

More than the shuttle was askew in NASA, as that kind of decision making indicates. Just in the shuttle itself, three different centers were involved in management. Their lines of reporting were complex by the time they consolidated at the upper two layers of the shuttle program. Warnings of the engineers at the working level about low-temperature dangers never reached the top in the case of Challenger. This was but one example of NASA's institutional setup overwhelming good engineering and simple, direct management structure. NASA was understandably protective of all its centers, for they held its reservoir of technical talent, hard to get and hard to keep. Head counts were rising, to be sure, but with support and overhead people. Then the centers themselves tended to run their own show without much interference from headquarters. They needed work to stay alive as NASA's budgets barely stayed even with inflation. So pieces of programs like the shuttle were divided like a pie among the centers, rather than concentrated at one.

More Embarrassment

After NASA had grounded the shuttle for two years for modifications, it resumed flights on a level of caution that was well warranted but too infrequent to serve its military, scientific, and commercial customers. At the same time it forged ahead on a manned space station with experiments and money from Europe, Japan, and Canada, despite warnings the program was floundering. Then another embarrassment raised new questions about NASA's ability to lead vanguard U.S. technology programs. While the shuttle put the $2.5 billion Hubble Space Telescope in orbit successfully in April 1990, astronomers found soon enough that its first images were blurred. When realignment commands from engineers on the ground failed to sharpen these images, NASA had to admit that the primary mirror was a microscopic 0.002 mm out of shape at its edge. After all the promotion of Hubble's ability in the vacuum of space to far outsee earth-based instruments, important parts of its imagery now were not better than or even up to terrestrial standards.

Most embarrassing of all was that another investigating board[5] found the error had probably occurred ten years earlier. Then a technician at Perkin-Elmer's Connecticut optics plant, later acquired by Hughes, had made a mistake in assembling an instrument, a reflective null corrector, to check the curvature of the primary mirror. Not only were there warnings in tests at the time, which were ignored, but the telescope was also assembled and flown without a system test. NASA chose to skip simple tests that would have caught the error. Instead it relied on the results of a single instrument to avoid running up costs of the behind-schedule telescope. Another inexplicable sidelight was that one of the company's most experienced optics experts chose to retire at that time and no one did anything to delay his leaving or take extra caution with less-experienced hands.

Not to condone the Hubble fiasco, but spaceflight, technology in general, is cut, try, make mistakes, fix, and finally succeed. Besides, some of Hubble's other instruments have performed brilliantly. And NASA plans a shuttle mission to fit Hubble with what amounts to eyeglasses. Nevertheless, Hubble is typical of the way the U.S. civil space program has evolved into something different from the way it started. In space science NASA had seized on very expensive spacecraft for one science or mission that crowded out others. Hubble flies yet one more red flag. A top industry engineer, who doesn't want to criticize NASA out loud, said privately that NASA has to figure out how to organize itself. "Look at the space telescope," he scoffed. "There was no integrator for that whole project. NASA bought solar arrays from British Aerospace and had vibration problems with them. It bought optics from Perkin-Elmer, and we all know what happened there. And the litany goes on." By integrator, he meant that not only must someone be in charge, but someone with the technical breadth to see that all the complex pieces fit and work together reliably. Either a prime contractor does this or the government acts as its own integrator. There were only associate contractors on Hubble, and the NASA center ostensibly in charge did not have full control.

"Space station is worse," he went on. "There are three centers involved, plus headquarters. There is a system integration contractor, but integration is not going very well because the contractor can't find out anything. NASA's got to figure out how to unify this loose confederation of centers and put somebody really in charge and accountable. These projects are so long, people know that the 'stuckee,' the guy in charge when the reckoning comes, will be somebody not even involved in the beginning."

Even fans of NASA fear that the space station—the centerpiece of American space technology for the near term—is headed down the same road to disaster. When it was first proposed, as a kind of international center in space to do all kinds of experimental work in materials and pharmaceutical processing, earth observations, space particle physics, and on and on, it was to cost $8 billion. Now the cost is estimated at $30 billion—and counting. Gene Levy, from his planetary exploration perspective, makes a shrewd observation that the space station is a contest between good choices—or perhaps indifferent choices—and better choices. Except to the taxpayer's pocketbook, the space station will do no harm and could even offer useful engineering experience in space in the daunting task of simply putting it together in orbit. The numbers keep changing, unfortunately, on how many hours of astronaut time floating free in space in pressure suits will be needed to do that and whether the human body is up to those demands.

"Space station," Levy said, "was going to have eight or nine attach points on the outside for payloads. There were going to be all kinds of important investigations going on. Now what? It's continually descoped, the payload attach points are all taken off. There's no support in the private sector for the materials processing. Almost no role is left for scientific investigations." An influential earth and planetary science society, the American Geophysical Union, has taken a formal position that the station's earth and space science capabilities don't justify its 29 percent share of NASA's R&D budget.

While the life science work might be important, the Soviets have been doing that in a space station put together a decade ago from more or less existing pieces. And they would like to tell us about the rigors for astronauts living for months in space, mental and physical, if we care to listen. Worse than that, in the 1970s the United States had its own space station, Skylab, created out of one of the Saturn booster stages left over when Apollo was canceled. Had someone thought about it at the time, Skylab could have been fitted with a kick motor to keep in in orbit and with adapters to add sections in space, just as the space station program today was originally planned. And it had a solar telescope on board. For that matter, the starving Soviet space apparatus has put its orbiting Mir space station on the block for something less than a billion dollars. Cancellation of the space station, named Freedom, while perhaps richly deserved, could precipitate an international incident because of the international commitments. Congress has been surprisingly tolerant of this exercise, although the House came close to cutting off funding in the spring of 1991. The advisory commission on the future of space, which reported to NASA and whose membership was partly drawn from space industry and government ranks, supported the idea of scaling back the space station, of development in a modular, evolutionary fashion. But scaling back narrows its usefulness. Less dependence on the shuttle was recommended, too, for the shuttle will have trouble supporting a space station.

For the more grandiose space missions the administration talks about, the shuttle is an even weaker support. The shuttle cannot haul enough payload into orbit or stay there long enough. So lunar colonization or manned Mars missions will meet an expensive new launcher and nuclear propulsion, both included in the fiscal 1993 budget. To its credit, the Bush administration, in the person of Vice President Dan Quayle and the national space council he heads, went to work on NASA early in the fiscal 1993 budget cycle. NASA's astronaut head, who had brought the shuttle back from the Challenger disaster, and the shuttle director left their jobs. Coming next: less grandiose spacecraft, fewer space spectacular and more workaday missions.

Far Out, but Stimulative

Another side to space colonization can be found in a building where the University of Arizona has leased office space for one of the centers of excellence that are becoming popular in government as a means of fostering technology. Under the formidable name of the NASA Space Engineering Research Center for Utilization of Local Planetary Resources, researchers are tackling an equally formidable obstacle to a manned Mars mission or a lunar colony. Allowing for something bigger and more reliable than the space shuttle to come along, the task of supplying a colony or a Mars expedition of any duration still is staggering. "Where's the water?" No such questions permitted on a Mars mission—unless a fossil ocean bed has water stored underneath. Like a pioneer family going into town once a year, dad can't get back home and then say, "Where are the nails?" Not only does it take very careful planning, but the rationale of the Arizona center is also correct: some of the supplies will have to be mined or manufactured on site in a near vacuum.

Whether or not the center ever solves such intractable technical quandaries, it will do at least a couple of practical things. NASA's grant of $7 million over five years pays a director, Terry Triffet, a small staff, and a few graduate students. The latter are not trivial. Students have avoided specializing in space science for lack of jobs after graduation. NASA is thus recognizing that it had better do something, and Centers of Excellence are one effort.

At one of the center's periodic progress update meetings, there was a sense of Apollo-revisited atmosphere amid awareness that the Apollo spirit is gone, that the days of the space spectacular are over. Space colonization, either as a goal or as part of a post–Cold War technology strategy, will have to be sold to the taxpayer as an investment, not an adventure or a national exercise in exploration or macho leadership. Far out as they may seem, there are payback possibilities on the moon. One is prospecting there for a helium lode. Helium in its ^{3}He isotope form could be used as the fusion fuel, if fusion ever is developed.

Next to the sun, the moon, without an atmosphere or a magnetosphere and exposed full blast to the solar wind for 4.5 billion years, is a helium orebody, if a few Apollo samples are representative. A few details need working out: practical fusion; then mining an estimated eight billion tons of lunar soil, extracting the ^{3}He from the tiny particles on which it is trapped, and getting the stuff back to earth. Shoveling around eight billion tons of lunar soil is equal to the largest mining operations on earth. Operating mining machinery in the airless, waterless, low-gravity moon is a staggering challenge, but technically demanding and thus stimulating to technologies beyond pure space.

An alternative, favored by another Johnson Space Center lunar science stalwart, Dave Criswell, is a solar power base on the moon to supply the earth. Criswell, now at the University of California at San Diego, noted how Peter Glaser's solar power satellite concept, once blessed with $30 million in NASA funding, has run into obstacles. Even with assembly in space, Criswell said, a solar power satellite is something like building the Grand Coulee dam in orbit, taking a Saturn Apollo-size payload launched every day for a year. Criswell thinks that lunar solar collector arrays are competitive economically, using lunar materials for construction and working out another detail: beaming microwave power to rectennas on earth without microwaving the people next door.

A National Technology Goal?

Far-fetched as all this may seem, it is the kind of technical challenge that made Apollo as stimulating to the nation as it was. The question remains whether this is near-term enough to serve as a successor to what has been the U.S. defense industrial technology policy. Possible, but it doesn't use enough people or industrial capacity if political make-work is the goal. Defense research and development alone is three times the present NASA $15 billion budget, which includes money for housekeeping and for salaries of its bureaucrats. Besides, the skills are not necessarily the same. Construed as a defense welfare program, space is not a great vehicle.

Space has the same ingredients as another big science spectacular: the $6- to $8-billion Superconducting Super Collider the federal government is building in Texas. This project needed $1.7 billion in foreign contributions, Japan as the target, to satisfy Congress. But Japan is shy about coming on board. As a Japanese official acidly defined an American Big Science project: one designed by American scientists to be built and run in the United States. When Congress cuts off the funding, it then becomes an international project.[6] Between the space station and the Super Collider (which Congress in mid 1992 threatened to kill), enough federal research money will be used up to leave many other university—and government—science projects starving. Besides just the money, there is opposition that the Super Collider basic research benefits are like those from space: too far away in time and doing nothing to redress U.S. industrial competitiveness. Objections also echo those about government aeronautics funding: benefit for too narrow a group.

A final word from Gene Levy: Federal policy makers have not capitalized on what space has to offer. "We have a commitment to human beings in space," he said, "which is wonderful. I like watching people float. I'd love to do it myself." Then there is a quasi-commitment to put people on the moon and on Mars. Both have slipped in schedule a decade or more. "We can certainly do it," he said, "but we haven't stepped back from the president's rhetoric—and I think the president's rhetoric was written by NASA—to ask ourselves what role the NASA program should play in serving the nation's needs."

Visionary goals are fine, but even the space fraternity is not all buying that. As the old commercial asked: where's the beef? Space indeed is worth its own keep if its goals are reasonable and articulated to win wide support. But it is part of an overall industrial and technology strategy, not one unto itself. "People in government are, in my opinion, prisoners of a system which historically has not been asked to deal with the set of problems that now is hurting the country," Ralph E. Gomory, president of the Alfred P. Sloan Foundation and a former senior for science and technology at IBM, told a Congressional hearing in the late 1980s. "We have problems of true internationalization and very real problems of manufacturing, but in Washington, money is still going toward space spectaculars. . . ." As the United States cannot continue to carry the burden of free-world defense while its allies appropriate its commercial markets, neither can the United States spend vast sums on long-term far-out research while short-changing near-term, more mundane industrial applications. Unlike the supersonic transport's definable market and payback, space has yet to define one. Spending $30 billion for space station spectaculars where astronauts test-ride exercise bikes to adapt to weightlessness won't show the return on investment that the 1990s will demand. With the 1991 recession, with the new cold war, with defense shrinking as a technology driver, technoindustrial policy must be pragmatic. Both the supersonic transport experience and space exploration are guidelines in what will work and what will not.

Revitalizing Industry

Dichotomy is a pretty good word for the way this country has approached both military and civil technology—what some define as industrial policy—over the last decade. Civilian applied research had tough sledding, owing to the misfortunes of the supersonic transport, breeder reactor, and space boosters and vehicles. Not unreasonably, the Reagan administration shrank from picking industrial or technical winners and losers. With a military spin on the same technology, the money came easier. Picking winners and losers survived as a catch phrase into the Bush years, as long as John Sununu as chief of staff led the White House Troika. The dichotomy? Even before defense spending slowed down in the second half of the 1980s, this attitude was slowly changing.

Both Congress enthusiastically and the administration reluctantly had recognized that the United States had an industrial dilemma. The Reagan White House convened a blue ribbon commission to study the problem, which ranges from trade, tax, and regulatory policy to technology. By the time George Bush was getting comfortable in the Oval Office, Congress mandated that federal departments take a look at critical technologies for the future, by inference as a focus for research by the federal government. While not a true industrial policy, the drift was in that direction.

Smack in the middle was the White House Office of Science and Technology policy in the person of William D. Phillips, who had a background in both academic and industrial research. At the White House science office, he became chairman of the National Critical Technologies Panel, and thus liaison between White House and this group convened to report biennially as mandated by federal law on the state of critical technologies. Because of the winners-and-losers implication, the committee's report had to be issued with a disclaimer of any White House endorsement.

With an unsympathetic White House staff on one side, Bill Phillips also had to defend against those who thought the government was moving far too slowly in these technologies, both military and civil. At one Washington meeting in the fall of 1991 he fended off questions from engineers and professionals: Was the White House serious about critical technologies when its own budget office was declaring it had no enthusiasm for simply getting up a list, let alone picking winners and losers from it? Well, Phillips answered, critical technologies lists can become an unwanted bible, an instrument for micromanagement, for bureaucracy to meddle in day-to-day program management. "For the administration to keep arms length from this," he said, "to utilize it as necessary but to realize that there are problems with it, is perfectly legitimate." Another questioner asked: Was White House aversion to the list being used for micromanagement a recipe from the White House for no action at all? Falling back on the long, convoluted process of consensus building in Washington, Phillips responded that, "in your organization I'm sure all is not sweetness and light."

From those questions it is obvious that industry, even Republican industry, had reservations about the lack of action from the Bush White House on technology enhancement, on research to help competitiveness now and not basic knowledge to help sometime in the future. Despite the resistance of the White House—especially from the former Bush chief of staff, John Sununu, who tended to fill the vacuum left by the president in running domestic policy—the federal government was gradually becoming more involved. Heeding intelligence reports from the industrial war, the Commerce Department, in the days of Thomas J. Murrin as deputy secretary in 1989 and 1990, started research centers aimed at the shop floor. These were half-measures for those in industry who would like to see a federal Department of Manufacturing, covering both military and commercial. Phillips predicted, and no doubt rightly, there would be nothing that drastic coming out of the Bush White House.

Manufacturing research by the Commerce Department came by virtue of an act of Congress. Commercial research in general is under the aegis of the Commerce Department's National Institute of Standards and Technology, what was the old Bureau of Standards until it was given a new charter in the Omnibus Trade and Competitiveness Act of 1988. The act itself was a signal that Congress, in the waning Reagan days, was getting more concerned about U.S. industrial competitiveness than the executive branch was, a tradition maintained by George Bush until he moved in the other direction with his ill-starred mission to Japan with the heads of the U.S. auto industry. Money for Commerce's new responsibilities was slow in coming, but to give Commerce credit, it asked for proposals as soon as it had the legal authority and selected three of thirty-six submitted for starting Regional Centers for the Transfer of Manufacturing Technology. They were the Cleveland Area Manufacturing Program, another Ohio effort to revitalize its industry, and those at the Rensselaer Polytechnic Institute and the University of South Carolina. Four more are planned or in existence, one being Michigan's state-industry consortium, the National Center for Manufacturing Sciences at Ann Arbor, which represents yet another innovative approach on the regional scene. This 150-member center was the focus of an Energy Department cooperative research agreement in 1992 for a wide range of advanced manufacturing research. For every two steps forward, there is one back, however. While the fiscal 1993 federal budget included $18 million to help support these centers, the catch was that no new ones open until a contracted-out study is completed of how best to leverage the federal investment in them.

Despite its professed interest, Congress, while authorizing $40 million for the manufacturing centers, appropriated only $7.5 million in the 1989 fiscal year. With that precedent, the Bush White House included no money at all to continue their operation in the budget that followed. Somehow Commerce squeezed out about $35 million for the centers in 1991, but as John W. Lyons, director of the Institute, said in irony: "We're working on half a billion and we're not quite there yet. But as everybody knows, the Bureau of Standards, now NIST, is a very high-leverage operation."

These centers have impressed participants, says Alden L. Bement, Jr., who follows this field as vice president for science and technology for TRW, Inc., a diversified manufacturer of high-tech military satellites and less high-tech compiler of consumer credit reports. Since they are not so much aimed at new technology as at disbursing existing technology, they have helped small firms. "Many times the problem with small companies is that they don't know what they don't know," he said. "Among some small firms we have found revolutionary changes not only in product design but especially in manufacturing." Outreach of the centers has affected their total business and enhanced export effectiveness as well.

Commerce has its own critical technologies list, and, according to Lyons, bears the blame for getting the whole process started in the federal government, back in 1986. Critical technologies was not the original name. Instead they were called emerging technologies, to distinguish between those new on the scene and those equally important ones already on the scene but mature. "What do you do with steel making," Lyons asked, "which presumably is still critical to our national success but hardly a newly emerging technology?" That question persists in bedeviling industrial and technology policy, for it is mixed into the notion of government ignoring sunset industries but fostering those classed as sunrise. Composite materials of carbon fiber and plastic developed for military aircraft may well replace steel in automobiles some day, but the United States may have taken a national risk by standing by while Japanese or European steel makers revitalized an old technology, took away American market share, and started an American Rust Belt. Still, when the United States raised protective barriers for steel, U.S. Steel used the breather and its capital to buy Marathon Oil, not for technology improvement.

Nevertheless, in terms of overall manufacturing research, the Bush fiscal 1993 budget can't be accused of meanness. While it claims over $1 billion is included for this work, two-thirds of that is defense, which pioneered government manufacturing technology programs. Nondefense-advanced manufacturing research is budgeted to top $300 million, a 27 percent increase over 1992, which is progress.

Industrial Policy or Industrial Welfare?

Questions notwithstanding, there is a modicum of good news. All the private and public critical technologies lists, even from Defense, have shown a gratifying convergence, and that itself is significant to Albert Narath, president of Sandia National Laboratories. For that reason, Narath adds: "The whole matter of picking winners and losers should not be viewed as too important." The winners, by inference, have already been picked by the time the question gets into the higher reaches of policy making. Next—and essential—is to apply them for launch into the marketplace before their technical half-life is over, supported by a governmental technology policy to link technology to product. Narath concedes, though, that there are legitimate worries about any national technology (or industrial) policy. "How do we assure that the government support of critical technologies doesn't simply turn into a sort of welfare system for favored industries, federal laboratories, and the scientific research establishment?"

As reiterated in both government and industry circles, Narath advocates exploitation of critical technologies as industry driven and with ultimate focus on product. "Industry input, consensus, and teamwork should be the principal determinant of critical technologies program management," he said, expressing one broad consensus: that industry, not government, has the prime responsibility to solve the industrial problems. To the Bush or Reagan White House, such technology policy definitions were thinly disguised hands reaching out from industry and into Uncle Sam's pocket, synonymous with government financing research and development that industry ought to pay for itself or government bailing out sick industry with taxpayer dollars. But government may well take a policy stand against subsidizing or outright bailing out a declining industry and yet be unwilling to stand by passively as an industry dies. As John Lyons asked: what do we do about steelmaking, just as a technology, let alone an industry?

Raising the Commerce Department to equal stature with State and Defense so that it can promote competitiveness was one idea proposed in Washington, no matter how remote the chances are of transferring a lot of money around in the government, not to mention status. Create a National Competitiveness Advisor equal to the National Security Advisor is another proposal. Then there are ideas for government technology development programs like the supersonic transport, one to develop a 300-mph train system by 2020, with a $1 billion federal commitment. Rail technology was a futile thrust in the 1970s defense winddown. yet the need is there, if not the market. Texas is planning a high-speed intrastate rail system, but is likely to deal with foreign contractors where rail technology has survived better than in the United States. Seattle is looking at foreign high-speed rail technology as a potential solution to congestion and auto exhaust fumes. More of self-interest are ideas to nurture communications, software, and computer industries and to develop a national manufacturing initiative for computer electronics.

An electronics technical group, the national affairs committee of the Institute of Electrical and Electronics Engineers, began supporting the idea of a government capital "bank" early on. Where some have proposed a reincarnation of the Federal Reconstruction Finance Corporation of the 1930s depression, the electronics engineers consider that unrealistic in today's deficit conditions. They are proposing a variation, something like the private-public housing financing agencies Fannie Mae and Freddie Mac or the agricultural Farmer Mac. These don't use government funds, but rather borrow and sell equity in the open market on their government cachet for lending out, in this case to private industry that can't raise capital otherwise. Where these schemes often founder is that quasi-government agencies—or the Federal government itself—cannot borrow in the capital markets even at rates moderately less than other corporations pay and lend at half that—Japanese rates—to high-risk startup industry. Low rates in themselves won't energize the American economy unless there is new technology that needs easy money to make it into the world marketplace. Still an attack on capital-formation problems of U.S. industry that avoids direct government funding or outright subsidy is worth exploring.

A government-financed analog of this idea won the endorsement of a committee of the broader-based National Academy of Sciences. Headed by a Carter Administration Defense Secretary, Harold Brown, the committee proposed a Civilian Technology Corporation launched with a one-time $5 billion federal appropriation, afterward self-supporting like the Export-Import Bank. While its Democratic origins won't win friends in a Republican administration, the general concept was picking up momentum as defense shrank and deserves a careful look.

Catching Reagan's Attention

Amid specific industry agitation came a more generic approach from those who worried about U.S. industry. Ronald Reagan himself appointed his presidential blue ribbon commission to address the competitiveness situation. Former White House Science Adviser Jay Keyworth, coincidentally, was one of its members, and so was the late Robert N. Noyce of Intel, one of the architects of the defense-funded electronic research organization that became Sematech. The competitiveness commission report came in 1985 but still makes pertinent reading. That is about all it did in the White House, for its first recommendation made it unacceptable there: create a new cabinet level Department of Science and Technology. On top of that, it also recommended a similar Department of Trade. One can hardly blame the administration. Two new cabinet secretaries represented a lot of new bureaucracy for just one report.

While the Department of Science and Technology did not get far enough for a definition of what would be included, a congressional study[1] laid out a prospective list:

Agency	FY1986 R&D funding (millions)
Department of Energy (defense)	$2,987
Department of Energy (general science)	683
Department of Energy (energy research)	2,119
National Aeronautics and Space Administration	3,956
National Science Foundation	1,518
National Institutes of Health	4,726
National Oceanic and Atmospheric Administration	170
Bureau of Mines	59
National Bureau of Standards (now NIST)	101
	$16,319

Hypothetical as this list may be, it also shows that such a department could have considerable funding clout. At the same time, it reflects Cold War defense dichotomy again. No Defense Department facilities, none of its laboratories, were included even on an informal list, though the Energy Department's defense work was. No doubt that was a nod to the mix of defense and nondefense work at nuclear weapons laboratories. Defense then was still considered a world to itself.

Of all the changes coming in the post–Cold War world, the demise of defense as an island will be fundamental. The Federal Technology Transfer Act, which mandated that all 700 government laboratories sensitize themselves to commercial potential, was an early manifestation of what will be more of a dual-use way of life. Useful as the technology transfer act may be as a facilitator, it also is a warning to the laboratories to contribute on a wider scale or die. Several reports on just the military laboratories have criticized wasteful and duplicatory programs, though their consolidation was an early phase in defense cutbacks. "It is time to stop studying this problem and start implementing significant reforms," an Institute for Defense Analyses report said. More strategic direction for the laboratories and less micromanagement is needed from the Defense Secretary's staff, for the laboratories still have a vital role to play, but the report added: "Budget and efficiency needs will rightly force the closure and consolidation of many laboratories . . . but closures, consolidations, and reorganizations will not solve underlying problems of poor management practices and inadequate central direction and leadership."[2]

Institutional Impediments

While the idea for a Department of Science and Technology commanded attention in the blue ribbon competitiveness report, there were much less exciting but perhaps more important recommendations. They were embarrassing to the administration rather than anathema. One was to do something about the federal deficit. This would mean strictly curbing spending growth and maintaining steady, noninflationary economic growth. Less competition for capital from the federal government would lessen inflationary pressures and lower capital costs for U.S. industry, the report said, repeating a familiar point. Since those recommendations were offered, Treasury bond and note auctions have been expanding inexorably with perhaps surprising tolerance in terms of interest rates from investors. Both debtor governments and debtor businesses like easy money, and a thousand blue ribbon commission recommendations aren't likely to break that addiction.[3]

Structural tax changes also were proposed. Again, some of these were familiar. More taxation of consumption, less of saving and investment; reducing variation in tax rates on different industries because of credit and depreciation; fuller deduction of venture capital losses—these were typical. Stable monetary policy to reduce interest rate fluctuations and inflation premiums, another recommendation, has since arrived. But the 1990–1991 recession contributed more than any administration policies. High capital costs and low returns from manufacturing investment still rank as frequently cited causes for American industry's loss of market share, simply because companies just get out of a line of business if the hurdle rate, the return it needs on a capital investment to cover its capital costs, is too high. Or, without financing at home, they sell their technology to Japan or, desperate for liquidity, sell the whole company to any European or Asian company that can afford it.

Besides capital and tax impediments, American industry also carries the burden of aggressive environmental regulation and of employee health care and retirement. A common reaction of the businessman in the street is to scoff at industrial policy, or whatever it is called, and complain about regulatory burdens as the biggest cross U.S. industry carries. Scott Crossfield, the X-15 test pilot who first flew to the edge of space, was also an abortive entrepreneur. His distillation of the experience: "We could give American industry a new lease on life by getting off its back. If I were to invent the better mousetrap that would make me a billionaire, I would go broke getting established in business. And I already have."

In 1961 or 1962, when he was working on the Apollo lunar landing program, his better mousetrap seemed in hand. "It dawned on me that this country was using up all its helium at one hell of a high rate," he recalled. "NASA at Cape Canaveral would use carloads of helium to pressure check a booster and then vent it to the atmosphere. Just throw it away . . . and it cost 3.5 cents a cubic foot." Government agencies controlled the price, supply, and contracting for helium as a protected resource. To work around that control, Crossfield developed a system to reclaim, scrub, and repressurize that helium, venting it to trailers and pumping it back into a storage bottle for reuse at a saving of millions of dollars.

Indeed the government was interested. After submitting proposals for two or three years, doing what Crossfield suspects was free research on requirements for such a system, he started a company called United States Helium. Buying helium from the Bureau of Mines, United States Helium bottled it, scrubbed it, and sold it to many government and industrial users. United States helium grew into a $3 million business, supporting its continuing proposal to go into the recapture and reuse business. "Without any overhead," Crossfield said, "we did pretty well for a couple of years. Then Linde and Kerr McGee decided to get this upstart out of their business. So they bought into the contracts for a couple of years." Crossfield was working for a contractor on the Apollo space program, skirting a conflict of interest and not able to give the business all the attention it needed. Selling off the company piece by piece ended, he said, with it going "belly up, taking every damn resource I had."

Most of Crossfield's problem, in retrospect, was due to government regulation. Besides having to pay cash for a carload of helium, he had to buy it through the General Services Administration, the government's housekeeping buyer, to sell it back to the government. "I was penalized if the delivery was late," he said, "but I never got a nickel when the check was late." Uncle Sam was slow to pay during the Vietnam War and six or eight months of waiting was typical.

"So I had to factor everything I did," he said. "The bank was sitting there taking 10 percent right off the top before I ever got the money from the government." Good idea, good product, all winding down to nothing. "That's what we are running into in spades today," Crossfield says with bitterness, "as people who are not used to the government try to go into the competitive market."

Specific complaints are common, but a Massachusetts Institute of Technology Commission on Industrial Productivity study[4] found little case for the thesis that environmental regulation had been a serious detriment across the board. Social purposes have a price and the key is to avoid inefficient regulation. "The Commission's sectoral studies revealed specific cases in which regulation had a serious impact on performance, but we did not detect a major effect across the board. . . . According to one report, prepared by the Congressional Budget Office, the overall effect of regulation has been slightly more negative in the U.S. economy than in Canada, West Germany, and Japan, but in no case has it been a major cause of inefficiency."

Economic generalizations like that irritate entrepreneurs trying to stay afloat in what seems, especially to a small business, a morass of federal and state directives on how to operate. Inefficiency, and lack of local coordination, are universally recognized, as in the Commerce Emerging Technologies report. Lengthy and expensive evaluations of new products for health, safety, and environmental impact are an area where the government could show some leadership, it suggested. As an industry engineer wryly described new product development in his company: "The first people consulted in development are not the engineers, but the lawyers." If social goals match the cost, nevertheless, regulation is not a total loss. Besides, regulation led American companies to be leaders in emission control, an export technology as the rest of the world begins to worry more and more about environmental consequences of manufacturing.

Another View of Government Help

Then there is the differing attitude toward government's relationship to industry represented by the Reagan commission on competitiveness and its successor, Council on Competitiveness. The latter is a private organization of 150 industry chief executive officers who pay membership fees of $5,000 to $15,000, augmented by advisers from universities, affiliates from technical societies and trade groups, and advisers from labor. Its first chairman was John A. Young, president of Hewlett Packard and chairman of the Reagan blue ribbon competitiveness commission. Erich Bloch, who shook up things at the National Science Foundation before he retired as director, is a staff member, and Tom Murrin, who tried to shake up things at the Commerce Department, is on the executive committee. Ford's John McTague is on the technical advisory committee. To the annoyance of the private organizers, Vice President Quayle appropriated the name for his White House group, which has tried to follow the Crossfield format of getting government off industry's back rather than deal with technical or industrial strategy. Whatever its longer-term success, it has mostly got into wrangles with Ralph Nader kinds of groups over whether it was, in fact, obstructing health and safety regulation and enforcement. Quayle's group did bring about a moratorium on new government regulation as a prelude to the 1992 Bush reelection campaign.

Retired Navy Admiral Bobby R. Inman, an important figure in the national intelligence agencies, is an adviser and headed a study on technology priorities for the private council. As Inman points out: "If you look at the period 1982 to 1988, we added 8.8 million new jobs in the aggregate in this country. No other society matches that. But if you look closer at those numbers, we lost 1.2 million jobs in manufacturing, and 400,000 in the extrative industries. We added 10.4 million in what we loosely call the service industries, all the way from those working in the fast food emporiums to the investment bankers. The average weekly wage of the 1.6 million jobs lost was $444. The average weekly wage of the 10.6 million new jobs, skewed by the investment bankers, was still $272. So the dramatic impact for many of our citizens who are working is already very apparent." But for burgeoning two-adult working households, the income shrinkage would have been steeper. Implications are obvious if this trend continues. America will trade more higher-paid manufacturing jobs for lower-paid service jobs. A Census Bureau report using 1990 population figures supported this concern. What it classed as low-wage workers rose from 12 percent of the full-time, year-round labor force in 1979 to 18 percent in 1990.

Technology policy rather than an industrial policy, as narrow a distinction as that sometimes is, was the approach taken by the private council. Hailing the White House Office of Science and Technology Policy's achievement in getting a critical technology list germinated in unfertile ground, the council urged the next step: implementation, as the federal lexicon puts it. That is, the council advocates more intense federal, and state, spending to develop these technologies.[5] Industry must lead, Inman agrees, espousing a common position in the business world, in areas such as quality improvement, innovation, and productivity. Government must lead where industry cannot, in getting down the cost of capital for one thing. "Council studies," Inman said, drawing from his role as an adviser for them, "found that where the cycle from research to product was short, U.S. companies tended to do very well. Where there had been sustained government funding, reliable funding, for a long period of time, U.S. industry tended to do well. Some exceptions occur, as in the chemical industry, which did well without government funding. Its culture had already established the need for long-term investment. Where there was shared mix of both government and private sector sustained funding, the United States did exceptionally well. Information technologies stood at the top of that list. If a product required long-term sustained [private] investment prior to getting into the market, we did not do well. If the industry had very shallow capitalization, it did not do well."

Technologists, defense and commercial, tend to agree, surprisingly, on what the critical technologies are for the future: microelectronics, information, materials, and the like. However, these lists are so generic that it might be hard to take exception with them. Commerce's John Lyons points out that there are seldom any surprises on such lists, nor should there be if the list makers know their jobs. "Generic technologies have a special blessing," he adds, like thin films that cut across various kinds of products. "These things are appropriate for a government posture, even in commercial technology areas. So the generic and enabling technologies that are very broad are in fact highly desirable for some of us." When the first Commerce list was unveiled at a press conference, a handful of reporters was expected. In the event, it was mobbed, even though few at that time knew what these lists were, let alone the technologies on them. "In retrospect," Lyons said at a 1991 meeting, "they were probably more interested in whether the government would specify them. We weren't sensitive to listing them in those days, but we are a good bit more sensitive to that today."

Critical technologies lists do have a drawback, one that justifies the White House arms-length attitude. Crystal balls, even consensus crystal balls, are murky. Charles Townes, a Nobel laureate for his work in developing the laser, relates a story of a committee of leading scientists assembled by President Franklin D. Roosevelt in the 1930s to advise the government what technologies were going to be important for the next several decades. What passed for a critical technologies list then included plant breeding, synthetic gasoline and rubber, and improved electrical machinery. Missing were antibiotics, even though penicillin was known then, nuclear energy, radar, rocket engines, jet propulsion, transistors and solid-state electronics, computers, biotechnology, and, not the least, the laser itself.

Another drawback exists as well. Without a product at the opposite end, do government critical technology lists make a difference? Barriers to commercialization are seldom technical anyway, but,

Lyons said, financial, trade, or legal issues. Still, Bobby Inman added, quite a few policy makers will ask: "Is there really a problem here? After all, aren't consumers doing very well with those wonderfully high-quality, low-priced foreign products? Jobs get lost in that proposition.

Policy makers aren't the only ones to ask that question. Isn't the United States doing a dumb thing by slapping restrictions on inexpensive, low-tech Asian electronics or tools or textiles in a misguided attempt to protect its own manufacturers? In other words, the economist's comparative advantage argument, to let the countries best able to produce an item take the market. Comparative advantage was at the root of another case made by two consultants asserting that the U.S. computer industry should forget about manufacturing. Just ship the specifications off to Asia, they contended, and concentrate on America's strength: software.[6] Responses came from all over, along the lines that such thinking has got this country into its present fix, where its largest companies seem to be turning into marketing organizations to sell products manufactured overseas. Besides, can U.S. software developers maintain their lead if they don't understand computing hardware intimately?

One of the responders was Daniel F. Burton, Jr., a political economist who is executive vice president of the private Council on Competitiveness and was the day-to-day director of its critical technologies report. Like manufacturers generally, Burton found the premise worrisome. Important as systems integration of software is, he said, this argument discounts the importance of manufacturing, assuming that "manufacturing is incidental to creating utility for customers. But manufacturing is often closely linked to product innovation and can have a tremendous impact on software and customer utility."

An echo in stiffer terms came from an outspoken industrialist, T. J. Rodgers, president of Cypress Semiconductor. Nevertheless, he agreed with the argument that U.S. computer and semiconductor companies need to shape up, as critic of what he calls huge U.S. dinosaur technology companies. Bridling at the notion of the United States abandoning computer manufacturing and "subcontracting everything to the Japanese except the creative process," he added: "To be competitive, the United States must manufacture with zero defects, develop critical products and technologies on short schedules, and aggressively market those products with impeccable service. . . . We cannot become a nation of 200 million programmers of Japanese computers, nor should we keep alive at all costs some of the dinosaurs just because they create manufacturing jobs."

Whether a group like the Council on Competitiveness can turn the Bush administration around, or perhaps its successor, is a question whose answer is yet to come. The impact from downsizing the military's role in the economy, added to the existing U.S. structural problems in manufacturing, will not be felt in full force until the middle of the decade. Not many politicians in either fold apprehend that there could be a national industrial decline, with or without a military winddown. Dan Burton thought that one of the Bush advisers, Richard Darman, "has been getting more flexible on technology and research stimulation. He's looking at a budget deficit." After all, the budget director is the first one to hear if federal tax revenues start to fall when jobs and manufacturing go away. Michael Boskin, Burton added, is a classical economist and just very suspicious about government intervention. And John Sununu seemed to call the domestic shots, before the jams he got into with his free-wheeling use of government cars and airplanes and his abrasiveness with Congress. Then the boom dropped because of White House staff fumbles that included a remarkable early detachment toward the 1991 recession and its sluggish aftermath. Despite Sununu, the White House science office critical technologies ideas had turned up as initiatives in the 1992 fiscal year Bush budget, and they gathered momentum a year later with Sununu gone, an important one aimed at enhancing biotechnology research.

Not a lot of dollars are needed by federal standards to get started on industrial technology, and it doesn't have to be new money. "Our report said it's a question of priorities," added Dan Burton of the Council on Competitiveness, "and it shouldn't take that much money. Let's just stretch a few of these national prestige projects, the space station or the super collider, so as not to dump in all money up front." Not all the advocates for industrial or technology policy necessarily agree, but it is pretty clear there are ways to get at the money problem without industry's hand digging into the federal subsidy pocket.

Alarm, but not Consensus

All kinds of Americans are deeply worried about the relative positions of all kinds of sectors of industry to Japan and the European community. But John Lyons, at the risk of flogging the obvious, pointed out that little consensus exists. "What we do about this situation from the policy level across the board, in say the administration and Congress, isn't clear," he said. "Those two bodies have not yet developed any kind of common position. I don't believe the think tanks and U.S. business have a common agenda yet either." Neither does industry unanimously believe it has a problem, and some industries do not. Drug companies disagreed with an assessment in the Commerce Emerging Technologies report, the updated 1990 version, that the United States was lagging behind Japan in life sciences research and development and farther behind in new products. Actually, the report meant medical instruments, not drugs or biotechnology. Ford's John McTague, makes a good statistical case that American automotive companies are putting as much into research and development as their Japanese competitors. Technology or research is not necessarily the cause of a lack of competitiveness.

Assessments of U.S. competitiveness are eye-catchers, like the Commerce or the private Council on Competitiveness studies. They are meant to be, because, as Lyons observed, they are vital to operating heads in government to justify programs. In that sense, they are similar to the Pentagon's annual appraisals of Soviet military power or budget statements to Congress. Depending on what programs the Pentagon was selling, they showed that the United States was lagging behind in fighter aircraft production, tank production, nuclear missiles, or research in various categories of new defense technologies. These are all reasonably accurate, but sometimes are subjective, self-serving, or not necessarily the whole story. As the Soviet upheaval demonstrated, the Pentagon comparisons did not express how drastically Soviet military research and weapons buying were sapping that nation's economy. Yet that was a collateral justification for one military weapons spending program or another.

Issues of this kind were hashed over, but not out, at a late summer of 1991 Washington meeting of the National Center for Advanced Technologies, called in Pentagon-like acronym NCAT. This organization claims a first in the critical technologies list business, as part of its campaign to develop strategic plans for preserving the U.S. lead in eleven specific areas. These are beamed toward aerospace and defense, given the center's origination with an aerospace trade organization, the Aerospace Industries Association. Detractors in the aerospace field itself say that the whole idea is just to keep defense contractors in the government trough. John Siwhart, heard from before in the first supersonic transport debacle, helped to get the organization started and, as its president, obviously did not agree. "We cover fourteen of the Defense Department's critical technologies," Swihart says, "and half, or a little more, of the Office of Science and Technology Policy list." So the center has an eye toward more generic technologies, especially dual use.

Because critical technologies has caught on as a buzzword in both industry and government, the meeting attracted a surprisingly large crowd for a technopolitical meeting—about 250 people, including nondefense company and the usual Japanese and European Community intelligence gatherers. As for consensus, there was some, but fairly general. Dan Burton of the Council on Competitiveness is skeptical about meetings, but this one turned out better than he expected. "Turnout was a good sign," he thought. "People are getting alert to the problem." Thinking is beginning to crystallize, as on simple but important ideas like industry, not government, taking the lead. Private grousing was common, as well as impatience at indifference to U.S. industry's decline, or at hostility from the White House, and to the slow reaction time of government generally. Industry is charged up about cycle time, that is, in making the trip from technology to new product far faster, as Japan makes it happen. Government, the government people on hand admitted, even those from the technology agencies, is too enmeshed in its own procedures and shibboleths to do anything incisive or useful about industrial stimulation or easing the shock of a shrinking defense industrial program.

Shrinking defense was motivating at least of a few of those there: finding some way to diversify out of defense—and, being defense contractors, to look for some sign of government direction. One of the women there was interested in materials that might have dual use and commercial fallout. But she brought up a familiar obstacle.

Defense, she complained, is the tail wagging the dog in its many controls over technologies. "In my trips to Europe," she said, "I've routinely been shown things in development over there that we can't even talk about here." She touched on a long-standing dispute between industry and the Pentagon over how tightly the U.S. should restrict exports of medium- and high-technology products as common as personal computers, a dispute lately complicated by the way Iraq bought weapons from the West for the Gulf War. The issue will resurface as military claims on technology subside but transfer of civil technology becomes a weapon in the new cold war. For example, a Silicon Valley entrepreneur was bitter over slow defense response to his new color television technology, which, for lack of domestic capital, he had to sell to a Japanese company. Dichotomy, indeed. Consensus on the policy level supported the industry-must-lead approach while the working level was looking for a signal from the federal government on how to proceed.

Swihart's critical technologies group has developed 10-year research plans for two technologies, advanced composite materials and rocket propulsion, and more are planned. Materials cut across industry boundaries. Rocket propulsion is strictly aerospace, and its customer is largely government. Promoting rocket propulsion research is the kind of platform plank that gives other industries the idea that the center is essentially protecting the aerospace money supply. Sure enough, the organization's goals call for keeping U.S. aerospace from losing its international technical and market lead, as so many other industries have. Pump priming with new money is not implicit; rather the goal is to maintain research and development in key aeronautical technologies, both industry and government financing, at levels comparable to what it has been. Rightly, the organization senses a waxing mood in both private and public sectors that aeronautics, with the end of the Cold War, is no longer a private American preserve. Still, its campaign moved slowly afterward, and ironically, with defense cuts, the aerospace organization, short on contributions from defense contractors, is looking to a broader audience to stay afloat.

When the Office of Science and Technology Policy published its committee's critical technologies list, Bill Phillips recalls a common reaction: "Oh, no, not another list of critical technologies. Let's get on with doing something useful." As the private Council for

Competitiveness asserts, the era of making lists of critical technologies is over and it is time to do something to exploit them. Swihart agrees with this proposition and that a broader industry front is necessary. The time has come for coalescence; still to be determined: coalesce about what?

One simple virtue of the half-dozen U.S. lists, private and public, is that there is congruence, which goes to the issue of gaining consensus. The same is true if analogous lists done by Japan and the European Community are added. "We can be reasonably confident," Phillips says, "that all the pertinent technologies are covered. The process of identifying critical technologies should not go on indefinitely. Rather efforts, henceforth should be directed toward policy issues and action items." In the federal bureaucracy, that means something will get done one of these days.

"Will It Be Easy? Hell, No!"

So here we are, the twentieth century almost gone; two world wars behind us; a long, silent, debilitating conflict ending in both disintegration and opportunity; an American industrial hegemony established and then successfully challenged; the world's most powerful military-industrial complex to be dismantled. Yes? No? Is world peace at hand? What do we do with it? Bill Phillips from the White House shook his head in amazement at the Soviet upheaval and said: "It's like Yorktown in the American Revolution, with the British bands playing *The World Turned Upside Down*. Military, economic, and technology policies that have served this country well for 50 years may not endure through the rest of the 1990s, not to mention the next century."

For the first fundamental change, there is the transition in store for a massive and pervasive military structure. Two scenarios are possible. Most likely are stiff cuts, at least 50 percent in weapons buying if not in the total defense budget before the 1990s are over, the product of push-and-shove between administration and Capitol Hill. The other, not the best bet but not such a bad one either, is that Congress will not drive defense funding down, that its talk of slashing defense is just that. Just as Congress can't seem to stop spending anywhere else, Congress might just abdicate on defense cuts. Based on the experience of the 1970s, transition is going to be painful and difficult. As euphoria over the peace dividend began to fade by mid-1992, just as it did as the 1970s wore on, some in Congress, some federal policy makers, and especially affected communities began to recognize the reality that defense dollars and jobs would have to be replaced or that they would be financially worse off with peace, not better. If Congress breaks into factions protecting local interests, that would leave onus for job and contract cuts on the executive branch while Congress picks and chooses what it will accept in individual districts. This would leave defense in limbo, as a welfare establishment, in the worst kind of industrial policy, one that does nothing for U.S. technological competitiveness or innovations in defense forces for post–Cold War security.

Defense will shrink. It already had in the early 1990s. But how much is not as crucial as how fast. Attrition, which the Pentagon hopes will take care of its personnel reductions, takes time. Better to let production contracts run out for weapons programs. Terminations for convenience of the government to meet budget goals leave the government reimbursing contractors for inventory and work in process that could otherwise become equipment in the field, a waste. On the whole, not a tidy picture.

Budget cut plans proliferated in Congress as the extent of the Soviet disintegration sank in. Representative Les Aspin, the Democratic chairman of the House Armed Services Committee, had one. So did Republican Senator John McCain, naval aviator prisoner of war in Vietnam and a member of the Senate Armed Services Committee. McCain's plan was significant, because it convinced the White House that the original Bush military cuts were too temperate to survive in Congress. Starting out as more drastic, McCain's ideas eventually approximated the White House changes. Yet the Senate itself, generally more conservative than the House on defense, voted in the spring of 1992 to cut off the B-2 bomber sooner and voted less for ballistic missile defense than the president wanted. Paradoxically the Senate then put money back for a Seawolf submarine. As noted earlier, shrinking defense will bring out the best and the worst in the American political system. A Connecticut congressional coalition saved Seawolf, at least temporarily, as well as jobs in home state shipyards, despite the administration's judgment that it was not in the national interest to keep the program alive. Service interests are caught up in the turmoil, too. Senator McCain's reelection campaign opponent, a former Air Force comptroller as a three-star general, accused McCain of protecting the Navy and knifing the B-2 bomber, the project of another service. Cerebral searching for a grand strategy for the new cold peace will collide with military and individual congressional campaigning to protect what amounts to the status quo.

Otherwise, Congress in the fiscal 1993 budget review kept surprisingly close to the White House budget. Accepting the reality of congressional posturing on defense cuts and of fighting retreats by the services and contractors, the Bush line in the sand at 30 percent still is wishful thinking for the longer term, given the financial straits of the United States. Lacking a clear and present danger, with annual interest on the national debt about equal to the defense budget, with federal entitlements for social services out of control, something will have to give. Defense is the most discretionary of spending and thus the most vulnerable. Besides, there is no magic number for a defense budget that ensures adequate defense. Not only does the budget depend on the national strategic plan, but combat spirit of the forces, leadership, and morale are intangibles, not directly purchased by dollars.

Whatever happens to the defense budget topline, more drastic shrinkage in weapons acquisition is in store, again based on the pattern of the 1970s. Career disruptions and community distress indeed will be the social and economic price of peace as in the 1970s. Federal retraining programs will spend money for displaced workers to make the transition, and some may even find new careers that way. In the 1970s, though, retraining was marginal and, sadly, it will be that way this time. Government funding for defense industry conversion is futile and should be a candidate for budget cutting, though there are well-intentioned proposals in Congress for conversion loans. More important to the national welfare is retaining a core defense industrial base. Dual-use and off-the-shelf buying have limits but are the only things in sight to ease that transition. Less customized weapons simply will have to be designed to whatever can be found in the industrial base that remains or in the commercial dual-use marketplace.

Research and development for defense is the sector to watch. Defense was 80 percent of all federal research in the mid-1950s, dropped to only half in 1965, and fell behind nondefense research at the end of the Carter administration in 1979 and 1980. Then it more than doubled with the Reagan defense buildup, from $12.1 billion in 1979 to $35.6 billion in 1986, or 68 percent of total federal R&D. Pentagon intentions to hold it constant in real terms—at about $36.5 billion in 1987 dollars—were carried into the fiscal 1993 budget, although the defense share of all federal R&D falls 61 percent.[1]

Good grounds exist for preserving defense research and development. Here lies the future. Protecting it in the 1970s paid off in the Gulf War. Whether R&D alone can do the job for the long term without a production base to prove out the technology has been questioned before. On the other side of the case is the fact that defense will continue to bulk large in federal research and development, perhaps better than half. As such it preserves the Cold War status quo, the idea that big-ticket, high-technology weapon systems will dominate U.S. defense strategy. Further, it implies that defense should stick to its last and not recognize commercial technology utilization. Focus is fine in research, but the new reality is that defense research may not survive without returning more benefits to the rest of society. The alternative case for big military R&D needs consideration: that the new cold war will need weapons like those the British once called cheap and cheerful fighters, not hundred-million-dollar stealth aircraft. In other words, even smaller fighting forces will need affordable equipment in hand, not technology stored in computer data banks.

How total military research and development can remain constant while operations, people, and weapons procurement go down is hard to comprehend—except that some of those research-and-development-only programs will slide into production. Total U.S. research and development—industrial, university, and government—is likely to fall as defense goes down, though nondefense R&D topped $20 billion for the first time in 1989 after some leaner years in the Reagan 1980s. But the point is that research funding is not at the top of the list of America's national problems. How that research dollar is spent is the issue. Federal research leans toward billion-dollar space stations and Super Colliders, fascinating programs that don't do anything near-term to arm U.S. industry for the new technological Cold War. Critical technologies initiatives are encouraging, and so is the way NASA is carrying out supersonic transport research. But industry obviously is worried, as the National Center for Advanced Technologies proposals show, that the U.S. research base might easily suffer with defense shrinkage.

More important than how much defense goes down is how stable settling out becomes. Shrunken 50 percent but with steady funding, with a coherent national strategy, the military can cut the pattern of its long-range research and development, weapons acquisition, and force structure to fit an effective defense posture. Wild swings in the defense budget will leave the military having trouble convincing people to commit to a military career and push decade-long development and industrial programs into chaos.

Granted the pain, adversity is good defense in one important sense. As crisis is forcing corporate boards to demand more from management, as it forces improvements in manufacturing and quality, as it forces more from the workforce, so it will for the military. From bloated support forces and perks to expensive and complex weapons, adversity will change the military for the better. For like corporations, the changes will spring from inside, from those who know where the fixes must come. The trick will be to get the benefits from shrinkage without harming the future defense posture of the United States and without a net loss in technology stimulation.

Building U.S. industrial policy in the Cold War era around defense, necessarily so, made for subtle, but more pervasive effects in industry than generally recognized. Hence, when this policy is supplanted—or simply abandoned—the impact will be enigmatic but penetrating. Coming at a time when U.S. industrial competitiveness in international markets is shaky, and all markets are becoming international, diminishing defense technical and funding stimulation to the economy will be felt. How much will depend on how government and industry respond to the change.

While some talk about a new Marshall Plan to rebuild Eastern Europe in a free-market image, using freed-up defense talent to do so, the United States is going broke. The ability of this country to lead the world in a recovery from the Cold War has been diminished by debt and industrial base decay. Stimulative defense spending is being withdrawn at a time the U.S. economy is worrying about deflation almost 50 years to the date Japan attacked Pearl Harbor, unleashing a half-century of inflation. Not only will the military be faced with a drop to a 1930s share of the U.S. economy, but Japan and Europe have reemerged as a new form of ally-cum-threat, an economic and industrial opponent and partner.

Absorbing the Influx

So to the second element: Is the rest of the nation's industrial economy ready for this? Not the least is its ability to put to work those squeezed out of defense. America's industrial tribulations have hardly been insidious, at least in the sense of unrecognized. Besides books and articles, congressional hearings and blue ribbon panels have warned of the trend for at least ten years. "We got a lot of price shocks in the oil crisis," Bobby Inman says from his Council on Competitiveness platform, "but even then we didn't get much outcry until there were gas lines." What will it take to get people alarmed about the state of American manufacturing, he wonders: massive losses of jobs, sagging standards of living, a stagnant economy? One reason for the absence of panic is that American industry is working on the problem. The question is whether it can do it alone in an interdependent world. Should government replace its role in defense technology with something else, something to help upgrade commercial competitiveness?

Remedies that look to government more often than not wind up with two bifurcations of the same solution: create a new agency or department charged with responsibility for science and technology, for manufacturing or for competitiveness, or put more money into research and development to bolster industrial competence.

New agencies were nonstarters in the Reagan-Bush era. They ought to be in any administration. Creation of yet one more government agency simply moves the question over to a new bunch in Washington, a green staff that must learn what it's all about. Politically adept, they leave the tough choices to the next group. Broad-scale throwing of federal research money at industrial enhancement is a similar way to avoid straining the imagination for true solutions.

Agreed that the federal government is not in the business of bailing out foundering industry, it is in no position either to stand idly by. Supply-siders to the contrary, government does have a vital and central role in dealing with the viability of American industry. No more is that exceptional than to say that government is responsible for the national defense or, indeed, the national welfare. Government's own vital interests are at stake in any frittering away of technical and industrial capability irrespective of who is at fault. Nor can the government overlook the structural strains of the total U.S. economy beyond defense, in manufacturing, in banking, in real estate, in the dangers of a service-centered economy without a core like manufacturing to buy its services. Besides, in its own self-interest, it cannot let the defense base evaporate.

Government's dilemma will be acute as it pulls back its defense commitments. Cutbacks will be disruptive to the national economy for a time, perhaps for years, Pentagon plan or no plan. Defense workers will sooner or later find work in commercial areas, at some greater or lesser degree of pain, at some greater or lesser salary. Contractors will, in the buzzword of the 1960s, redeploy their assets, even if that is through liquidation and reinvestment. Whether laissez-faire can work through transition by itself is an open question, but a question government has to be prepared for if the answer turns out to be no.

What Washington needs most of all is a practical strategy, of when to get out of the way of industry and when to step in, a coherent strategy for dealing with things if they go wrong. Practical means a plan that does not require spending of vast new tax dollars or filling new corridors with bureaucrats. Federal bureaucrats, when they do get in, are used to running the show, issuing regulations, making the plans. That won't work in the coming transition. As industry has found out, teams—small, dedicated, and talented—will have to do the job.

Government doesn't need a new department or agency. Rather the old ones need a clear, simple, reasonable charter. Some possibilities:

• The Commerce Department, which historically has been charged with oversight of the nation's business. While Commerce is regarded on occasion as a joke department, it already has the responsibility for the industrial welfare. Commerce has the people and at least the nucleus of a commercial technology research organization—the Bureau of Standards reincarnated by Congress as the National Institute of Standards and Technology—to deal with industrial technology. And it does have an Advanced Technology Program also as a nucleus for a civilian research and development arm. While the program originally was not as fulsomely funded as Congress envisioned in the act to create it, the fiscal 1993 budget asked for $68 million to support government-industry sharing grants for research in critical technologies. As industry recognizes, small can be beautiful and government for these puts up 90 cents for every $1 from industry.

• The White House Office of Science and Technology Policy. As a staff organization, this office does not have the operating experience of government departments. Nevertheless, absent an outright government commitment to civilian research, OSTP could coordinate and articulate a national strategy for technological competitiveness. With a White House mandate, it could drive the Federal Coordinating Committee for Science and Technology process to get government on the same track with industry and academia.

• The National Science Foundation. With a mandate as an independent agency to fund nonmilitary research and, especially, to support the university research apparatus, the National Science Foundation has shown sporadic insight into the technology and educational foundations of industrial leadership. If national strategy centers on education as an industrial solution, on coordinating government-university-industry research, the science foundation has the background for the job. The real strength of America is this triple research team. Still, each arm wastes too much effort on suspicion of the others, and research and technology need integration. Government can't run the show, but it can push things along the way it wants—if it chooses.

• The Defense and Energy Departments. Let military requirements continue to drive government research and development, with an expanded mandate for technology transfer to industry. Alternatively, give the consolidated Defense laboratories a civilian mission.

Then there are variations:

• A manufacturing research extension service, similar to the agricultural extension service that, working with land-grant aggie universities, brought science and technology to the American small farmer. Commerce with its Advanced Technology Program is approaching such a system and the idea is popular with some in industry.

• A national laboratory along the lines of the Canadian National Research Council, populated in whole or in part with researchers, not contract managers, coming out of Defense laboratories, and open to all.

• An apostle for industry revitalization on the White House staff. For that to happen, the president himself must understand the situation. Airline deregulation came about because a White House advocate did the spade work and the rest of the White House from the top down bought into the idea.

• Government-created or -funded banks: for research, for Reconstruction Finance Corporation–style bailouts, to help defense contractors to convert; or microbanks for small or startup enterprises, something states or cities are starting to do on their own. All have their points, if Congress and the taxpayer agree the government has a good chance to get its money back.

• Regional consortiums of states and their industries and universities. Bubble-up from the shop floor rather than trickle-down of federal omniscience is attractive. Industry might well want to give up on Washington's ever understanding the problem and move where the moving is easier.

Of all these factors, the last is the sleeper. In the states like Texas or Ohio or Pennsylvania that are putting together their own regional consortiums and development strategies, there is skepticism that the federal government understands enough about what is going on at the grass roots to know what to do next, as defense comes down and the new cold peace arrives. Regional consortiums are not going to wait to see whether the federal government finds out. They arc going to make their own industrial policies. For they will cut their own deals, with other regions and with industry and governments overseas that have a better grasp of what is going on in the metamorphoses and internationalizing of industry than the U.S. government. For the federal executive branch, the best response would be the White House coordinator, one quick enough to stop departing industries.

Tapping the Pentagon

Along with ideas for new departments of science or of technology, trade, or manufacturing have come proposals for giving the Defense Advanced Research Projects Agency, with its experience in dual-use technology, a civilian mission. Or, as Senator John Glenn proposed, create a civilian version of DARPA. Eric Bloch, the former director of the National Science Foundation and a believer in a tempered government role in research and development, doubts whether a DARPA concept would work.

"We have plenty of bureaucracy already," Bloch says. "It would be a new structure, and I don't believe that new organizations and structures are the solution to the problem." DARPA is focused on a single customer, defense, and links military requirements to industry. A government agency cannot serve as that link in the more diffuse commercial arena.[2]

True, Bloch no doubt looked at the creation of a new agency from the viewpoint of an established—and funded—government bureaucracy. The old competitors are not likely to welcome a new one for space, attention, and money. Yet his point is valid, whatever his institution bias, if any. At the same forum, Robert Stratton of Texas Instruments said: "Clearly the competitiveness problems of American industry are, to a large degree, the result of inept management." This is a way of saying U.S. business and industry needs to clean up its own mess. As for new departments, one only has to look at NASA, which changed from an innovative and gung-ho science and technology team in its early days to a bureaucracy absorbed in keeping its institutional infrastructure alive and funded. Government bureaus sooner or later become slaves to their own agendas.

For that matter, creation of a cabinet-level Department of Science is hardly a new idea. The proposal arose not long after World War II, led by the late Hubert Humphrey, then a senator. In a salute to the dictum that the more things change, the more they stay the same, the arguments for such a department centered around increased stature of a cabinet level secretary and elimination of duplication. Scientists themselves opposed the idea then for various reasons. Basic research might lose out in the intramural competition, not to mention the competition with established departments, such as Defense and Agriculture.[3]

While there are objections, the idea of giving the Pentagon or military laboratories a commercial mission has a case from pure expediency. Enmeshed as they are in shelves full of regulations and auditors who know every paragraph, military research and development managers are still the most skilled of any in government in running research jobs. When it wants to, when it is allowed to, the Pentagon can be fast and business-like. That was then, though. Too many companies now don't want to do business with the Pentagon because its stringent rules make transaction costs too high.

Senator Glenn himself doesn't care to talk about a civilian DARPA now. However, Bobby Inman's Carnegie Commission committee has come out in favor of it. Arguments for a civilian counterpart, perhaps a Comsat-like corporation outside the civil service—or for moving DARPA out of the Defense Department to make it a National Advanced Projects Research Agency—are those Eric Bloch raised earlier: DARPA's success in sponsoring development of generic military technologies, which have had commercial fallout, its knowhow in managing innovation, and its availability as a fine role model. DARPA itself warns against adapting the agency as a model. Not only does DARPA have a committed military customer with money, but it also has a clear focus for applying its technology and a pertinent technical core competence—things lacking one way or another in the commercial marketplace.

Behind the concept for a civilian technology agency outside the civil service lies a realization. While those who have no ideological objection to or support outright a government-led or government-backed technology or industrial strategy per se if it works, they, like DARPA's former director Craig Fields, recognize the limits of what government can or cannot do. Unfortunately, the American federal government has simply grown too fat, too slow, and hence too indecisive and ineffectual to do anything fast enough to lead in a competitive arena.

Take this question put to Jacques Gansler by a listener, just after Gansler, once part of the Pentagon,had sounded off once again on critical technologies and how to foster them: "We're getting smart cars, smart factories are possible; we already have smart weapons. All of them seem likelier than getting a smart government, able to move nimbly, to respond to new conditions with quick and accurate changes in policy, to transcend parochial interests of individual agencies or individual congressmen." In other words, who needs government in developing critical technologies? Except for government to do a better job in what already is its responsibility: in education, regulation, or monetary and taxing policies. In the larger sense, should government try to help industry if its own management created the fix it is in?

Faced with the laughter and applause the question stirred, Gansler said he found it hard to separate government's role in regulation or tax policy or education from its role as technology advocate, noting: "Can the government move rapidly? Categorically, no. If it did, it would probably swing back and forth even faster with no forward progress." The best he hopes for is a government gradually moving from a very negative position that is now primarily one of regulation and lack of trust. If the American government functions solely as an industrial regulator and other governments as industrial facilitators, it should come as no surprise that this country has an industrial shortfall.

To the suggestion that government simply get out of the way and let industry do things properly, Gansler demurred: "Government is just too much involved in critical sectors of the economy." Especially is this true in research and development, if critical technologies are pursued as dual use. Government invests $38 billion in defense R&D, which supports one out of every three or four scientists and engineers, perhaps one in every ten factory workers, sometimes 20 percent of capital investment, and it pervades U.S. industry, particularly critical sectors, like electronics and aerospace. "It's obsolete to say the government ought to just get out. The government's not getting out." Government is in: in banking, in taxes, trade, regulation, in R&D. Government prohibits factories from integrating; R&D from being dual use; the Defense Department from buying commercial parts. Regulations obviously will have to change. "Is it going to be easy?" he said. "Hell, no. It will be very difficult, even more so if government doesn't recognize the problem." Industry must lead the way, but government can't tie both hands of industry and then expect it to succeed.

Going for the Splashy Breakthrough

Big government likes public relations glitzy, sweeping solutions to national problems. Here the solutions may be far more humdrum. In fact, there have been scores of solutions offered to the industrial competitiveness, though not so many to dealing with the effects of defense shrinkage. The difficulty is getting even easy solutions before the decision makers and then, beyond that, getting something moving. Half the battle would be over if the White House cared and decided to set out a strategy for dealing with the industrial and defense posture of this country in relation to technology in what most agree will be a much-altered world. Listen to Tom Murrin again:

"After two years in D.C., though I visited here many, many times—I don't know why I had to come for two years to learn this—I had no appreciation of the extraordinary difficulty of coming to consensus on any subject in Washington. It makes one judge on occasion it is almost literally impossible." Why that is so was driven home to Murrin when he took the chairmanship of the then-new technology and industry committee of the FCCSET, the Federal Coordinating Council on Science, Engineering, and Technology. His associates at the Commerce Department began mulling over how many federal agencies should be invited to its first meeting. "Very quickly," he said, "we made a list of at least half a dozen of those who are very much involved in technology. For a lot of other reasons, perhaps protocol or politics or whatever, we invited nineteen."

Two aspects of the meeting struck Murrin besides the mob scene. Most of the nineteen participants, generally number two or number three from their respective agencies, had never met each other before. Even more sobering, the following week Murrin got five irate telephone calls or letters from people in other agencies who wanted to be invited. So the next meeting had twenty-five people. "In a sense, wonderful," Murrin said, "A lot of brain power and democracy in action. But to come to consensus with twenty-five people who don't know each other is a hell of a challenge."

That puts into some perspective the caution White House science aide Bill Phillips has about what comes next with critical technologies and his soft answer to those who think the federal government has done little to take the next step. The FCCSET technology coordinating committee has been revived once more by D. Allan Bromley, heading the Office of Science and Technology Policy. Since 1989, the science adviser's office has been attempting to integrate various technology programs. This is a way of getting around using the awful phrase "picking winners and losers." President Bush's budget message in 1990 included the FCCSET global change program, which is interesting scientifically but doesn't do much for industrial competitiveness or defense redirection. The 1991 budget message included two more pertinent initiatives for industry: high-performance computing and communications, and improvement in math and science education.

"Two major cross-cuts are going on today," Phillips said in the fall of 1991, "and are right in the middle of the interagency correlation process." Their success was uncertain then, but in the January 1992 budget message they made it: one in advanced materials and processing and another in biotechnology. A $162 million, 10 percent increase in 1992 materials research money involving ten federal agencies through the FCCSET coordinating process was aimed at, among other things, improving the cooperation between universities, government laboratories, and industry, and moving technology from the laboratory into applications. Defense has $30 million of a $1.8 billion materials and processing program, but is the only major agency whose financing in this area is falling. Biotechnology would get 7 percent more, $271 million, in a twelve-agency FCC-SET effort that ranges from cost-effective biofuels at the Department of Energy to biological sensor technology funded by the Pentagon for threat detection at sea and drug interdiction. Vice President Dan Quayle's Council on Competitiveness figures in this area, too, in ideas to ease regulatory burdens.

While these are not monumental increases, nondefense research is on the right track. Manufacturing is an important facet and the principle of industry leadership is recognized. Then there is integration through the FCCSET process, no small achievement. As Tom Murrin recalled from his Commerce Department tour, consensus is an agonizing process inside the Washington beltway. But the FCC-SET process is useful and an example of the kind of role the presidential science office can play.

Murrin's other, perhaps surprising, insight is the limited impact of industry on government, hence the significance of recognizing industry leadership. "I used to have delusions of grandeur when they put me on a Westinghouse Gulfstream in Pittsburgh and flew me to D.C., there to be met by a limousine, resplendent with global advisers. We'd go to visit some senator or congressman, and it was all very exciting." Afterward, the group would get together and Murrin would ask, "How did it go?" "Oh, it was absolutely magnificent, Murrin. You were spellbinding. The senator took notes. He asked for copies. This will have tremendous impact," So Murrin would think, "Isn't it good I came. I saved the country." Then Murrin went to Washington as deputy secretary of Commerce to find that the typical audience on Capitol Hill is seven minutes—surrounded by 15 minutes waiting and 15 minutes finding a way out of the building. So it seems longer. "They smile at everybody," he said. "What you mostly find is that preceding you that same day and following you that same day is a succession of visitors who want to talk about owls and shrimp or greenhouse effects, all kinds of wondrous things that make your parochial interests seem like a pain."

Business people should try to influence Congress, but should not be consumed with their own self-interest and the issue at hand. A congressman or senator's main interest is how this business issue, whatever it is, affects the prospects for reelection. "That's very logical and very necessary," Murrin said, "and the agony is business can often package proposals so that they serve that deep need."

Keep It Simple, Uncle Sam

To deal with the challenges of drastically reshaping the defense industrial base in an uncertain industrial competitive climate needs simple approaches, not complex ones. The federal government doesn't have to wreak magic or bestow tons of money; it just has to understand the relationship of science, technology, and industry, act as if it is important, and lubricate the gears. Al Narath of Sandia puts the very sensitive relationship of science and technology well:

"Here is an area where there continues to be tremendous confusion and unnecessary debate. Science is the essential foundation of technology advancement. Without science we can't have technology. Science without simultaneous technology advancements is a foundation without superstructure. We need to protect our investments in the basic sciences. On the other hand, dramatically increasing the federal investment in basic science alone is not the way to foster critical technologies. Science and technology need to move as they always have in the past, in parallel." In other words, research per se is not the problem.

Science is available to all. So in itself it is not a differentiating strength in global competition. So often noted as the thing Japan does so well and America not so well is the ability to channel scientific knowledge into applications, rapidly and effectively. Science provides national competitive advantage only to the extent it is closely coupled to industrial product development cycles. Development cycles in the United States are too long, but industry, not government, has to solve that problem—as it is starting to do.

Government and the science and technology apparatus must collapse the barriers between research, development, and application. "Unfortunately," Narath recognizes, "there has been a long-standing bias within our scientific and technical community that views manufacturing methods and process technologies as too applied and therefore lacking in intellectual challenge. Very encouraging to me is [the fact] that national critical technologies lists place heavy emphasis on manufacturing." As for the government role: "Even to me, even as a techie, it is clear that the most important role government has to play is to create an appropriate business climate, a supportive climate. It's not a technology issue, but it's very important and has to come first."

Government does have a nucleus for fostering industrial technology in the Commerce Department's Advanced Technology Program. But the Bush White House initially no more than tolerated the activity, though, as noted earlier, its grant program began to grow in 1992, with more to come in 1993. Still, these programs individually are small—$1 to $3 million—alongside DARPA's $100 million to $1 billion programs. So industry doubts whether there is enough muscle in Commerce to do the job. Not many in industry or technology would quarrel with the notion that the federal government must coordinate better federally funded R&D at the national level.

For a simple way for government to stimulate industrial innovation, there is Small Business Innovation Research grant. While its funding of $450 million was not small, it does tap the imagination of the small, fast-moving entrepreneur. And the critics at its inception have been won over, at least to the point of not fighting an increase to $900 million.

In terms of preparing for the defense fallout, for a new industrial technology era, George Bush leaves a leadership vacuum because of his inability to convince his followers that he knows how to foster U.S. industrial competitiveness, phase out big defense, or deal with the economy in general. Wisely, the president shied away from government subsidy to industry for product development, like the round-one supersonic transport. But he has bought the big-science extravaganzas such as the space station and the Super Collider without getting them launched unequivocally or with public enthusiasm while giving a certain amount of lip service to the nitty-gritty kinds of manufacturing research that could help in the trade wars. Technology policy in the Bush White House, outside of the science adviser's office, was ad hoc through his first three years, and near-bankrupt. Other than calling on the Federal Reserve to lower interest rates one more time and on Congress to cut the capital gains tax, the president had shown little understanding of the deeper structural ills of the American industrial economy and what a defense standdown will do to it.

With John Sununu gone, things changed. Clearly the White House Science Adviser's office made progress with its critical technologies advocacy and interagency coordination of R&D. Not only was there $150 million more in the fiscal 1993 budget for high-performance computing, there were the increases in advanced manufacturing research and biotechnology and in other work, in all exceeding $4 billion. Yet it is best to view these with caution. In the early days of the Strategic Defense Initiative, the Reagan Star Wars missile defense program, the impression was left of a new, multibillion-dollar budget effort. In fact, the budget had simply aggregated all the money the services were spending for missile defense research and stuck it under a new heading. So does the fiscal 1993 budget aggregate research going on at a dozen or more federal agencies, with a sizable defense component. Caution to be sure, but also credit; coordination of these research efforts through the Science Adviser's integration system is a definite plus. Star Wars eventually became a real pioneering research program in its own right. Industrial competitiveness may well follow.

Congress has done a bit better. The technology transfer act was foresighted, and so was the trade bill that started the industrial research mill running, albeit slowly, at Commerce. Congress will make things chaotic for defense, but it may well be quicker than the administration to seize on the offsets in commercializing technology.

Who Has the Key?

In the early 1980s, when the Reagan recession and Japanese competition had thoroughly rocked U.S. industry, Professor Robert B.

Reich of Harvard enunciated the basis for an American industrial policy: "Real social costs are generated when industries decline." In other words, if American steel companies founder, thousands of workers lose jobs. Finding and retraining for new ones is expensive and slow. "Government," he added, "must respond to the competitive strategies of other advanced industrial nations." First into the market wins, as witness the U.S. lead in aircraft, computers, antibiotics, and satellite technologies where government's hand was helping. In particular, industrial policy must be coherent with government's tax and regulatory roles and must be based on broad consensus.[4] Later in the decade, Professor Reich turned less sanguine about industrial policy, recognizing that this country indeed fragmented those tending to serve special rather than national interests, some winners, some losers.[5]

Then, at the end of the decade, as defense began to come down in the aftermath of the Cold War, Pietro S. Nivola, an associate professor of political science at the University of Vermont, argued the winners-and-losers case, that industrial policy builds a political pork barrel, not a better outlook for American industry. He scoffed at the idea that success always followed either lavish industrial subsidies in Europe or Japan's vaunted government-industry collaboration; Japan's industry may well succeed despite, not because of, government's participation. He questioned whether matching grants to commercial industry or tax credits do more than have government pick up the tab for what companies ought to pay for themselves. And as for an aggressive trade policy to support U.S. industry, he concluded that it is about as coherent, coordinated, and active as could be expected. Yet surprisingly, or perhaps not so surprisingly, he also argued that, given certain assumptions, lodging U.S. technology policy in the military, with DARPA, is less duplicatory than funding separate military and commercial research and no more inefficient or confusing than the civilian agencies in America's overseas competitors.[6]

Both views have points, National government is in too deep to get out. Unfortunately, as George Kozmetsky fears, the federal government is running way behind. Too often it doesn't know how to transfer technology, how its laboratories should spin off defense technology or make dual-use technology into the smart civil infrastructure the United States needs for the twenty-first century.

"When Reagan and Bush declared there would be no industrial policy at the federal level," Kozmetsky said, "in effect they passed it back to the states." As a result states and cities had to start worrying about industrial competitiveness, which they did in ways to suit their own special needs. Kozmetsky's organization hit on the concept of technology venturing, to develop next-generation industry and promote Austin and Texas. Now the Japanese model is attracting local attention: regional complexes of university and research institutes, like the one planned around Osaka; aerospace, electronics, and biotechnology manufacturing malls; and transportation infrastructures. Cities, states, and regions will be forced to fill the vacuum left by the federal government in picking technical winners and losers. They will create their own, in Kozmetsky's words, smart infrastructures: a federal laboratory, if one is handy; an aggressive research and entrepreneurial university; local technical and business talent; and local capital. These will make their own industrial policies, networking, and joint venturing at home or abroad to do so.

This country must have an industrial policy. Even a choice not to have an industrial policy is a policy, if industrial policy is defined as bankrolling American industry. Letting laissez-faire do its best is a viable policy, as long as government has a plan to activate if laisez-faire stumbles. Better, though, is a technoindustrial policy aimed not at competitiveness per se, but at taking the new knowledge America is good at finding and seeing that American companies get its applications into the market expeditiously. This does not have to be a mammoth federal effort and must not be a case of the federal government micromanaging technology. A national laboratory or, given the size of this country, a set of regional federal laboratories open to use by gestating industry is an option. Small or startup companies need help with innovation and development, with someone they trust. A research bank is a reasonable concept for financing this kind of small-company innovation, in place of federal grants.

Washington recognizes belatedly that it has to do something in technological competition. More critical technologies research and development indeed is part of the fiscal 1993 budget. Yet there is a sense of lip service to technologies like manufacturing that the federal government is not familiar or comfortable with. There is a lesson in buckyballs, the carbon 60 molecule synthesized at the University of Arizona in the early 1990s. Not only was the cost of the work minuscule, but the federal government contributed not a penny. Money came from the state, in salaries for the workers and for equipment that was shared. Naturally there is federal research now, for one in using carbon 60 in an ion engine for spacecraft.

As defense winds down, the federal government will lose its ability to drive the direction of research and development. Ideally, the federal government, which controls trade and tax policy, should have an integrated technoindustrial policy or strategy or goal. Industrial policy, if it means bailouts and support of sunset industry, is a poor place for federal dollars. Let these industries set up their own cooperative insurance funds, if that is what is needed. Federal dollars, as the transistor and the buckyball show, don't invent things. But, as defense amply demonstrated, federal dollars do develop technology. That role will be just as vital in the new cold war, if the federal government acts with more conviction and cooperation and less micromanagement of the technology triumvirate of the next century. Otherwise, cities or regions will cut their own national or international deals and perfect their own technology venturing.

Notes

Chapter 1

1. *Budget of the United States for Fiscal 1993,* U.S. Government Printing Office, Washington, D.C., Part One, p. 96.
2. Congress of the United States, Congressional Budget Office, *The Economic Effects of Reduced Defense Spending,* Washington, D.C., February 1992, p. 27.
3. U.S. Congress Office of Technology Assessment, *Holding the Edge: Maintaining the Defense Technology Base,* OTA-ISC-420, U.S. Government Printing Office, Washington, D.C., April 1989.
4. William H. Gregory, *The Defense Procurement Mess, a Twentieth Century Fund Essay,* Lexington Books, Lexington, MA, 1989.
5. Ibid.

Chapter 2

1. Jeff Gauger, *Omaha World-Herald,* April 28, 1991, p. 1.
2. *Wall Street Journal,* Nov. 6, 1991, p. A-12.

Chapter 3

1. *Council of Economic Advisers Annual report,* Washington, D.C., 1970, p. 31.
2. *Council of Economic Advisers Annual Report,* Washington, D.C., 1971, pp. 24–27.
3. *Survey of Current Business,* U.S. Commerce Department, Washington, D.C., January 1992, p. 47.
4. Gary Putka, "Textron Defends $600 Million Acquisition of Cessna," *Wall Street Journal,* Jan. 23, 1992, p. B6.
5. William H. Gregory, *The Defense Procurement Mess,* Lexington Books, Lexington, MA, 1989.
6. U.S. Congress, Office of Technology Assessment, *Holding the Edge: Maintaining the Defense Technology Base,* 1989, p. 5.
7. *Integrating Commercial and Military Technologies for National Strength: Re-*

port of the CSIS Steering Committee on Security and Technology. The Center for Strategic & International Studies, Washington, D.C., 1991, pp. 20–23.

8. Paul H. Richanbach et al., *The Future of Military R&D: Towards a Flexible Acquisition Strategy,* Institute for Defense Analyses, Alexandria, VA, July 1990.

Chapter 4

1. James S. Coles, ed., *Improving American Innovation: The Role of Industry in Innovation, in Technological Innovation in the '80s,* The American Assembly, Columbia University, published by Prentice-Hall, Englewood Cliffs, NJ, 1984, p. 58.

2. "Corporate Giants and Global Competition: Dynamos or Dinosaurs, "Technology Policy Task Force of the House Committee on Science, Space, and Technology, 1987.

3. Stephen S. Cohen and John Zysman, *Manufacturing Matters: The Myth of the Post-industrial Economy,* Basic Books, New York, 1987.

4. Oddly enough, concurrency was the magic word in military development of the intercontinental ballistic missile in the 1950s. While the general idea was the same, develop key elements like warhead, rocket motor, and electronic guidance simultaneously, these were infinitely more challenging than the technology in commercial projects. When the inevitable delays hit one, overall schedules were disrupted and costs soared. So concurrency got a bad name in defense.

5. Paul B. Carroll in the *Wall Street Journal,* April 15, 1991, p. A-9C.

6. Richard W. Stevenson, "Breathing Easier at McDonnell Douglas," *New York Times,* Sept. 29, 1991, p. 6-F.

Chapter 5

1. Stephen R. Chitwood, ed., *Economic Policies for National Strength: The Quest for Sustained Growth and Stability,* Industrial College of the Armed Forces, Washington, D.C., 1968, p. 144.

2. An excellent account of the government's role in computer development is in *Creating the Computer,* by Kenneth Flamm, The Brookings Institution, Washington, D.C., 1988.

3. Much of this history is drawn from Albert D. Wheelon, "The Rocky Road to Communication Satellites," the Von Karman lecture at the American Institute of Aeronautics and Astronautics Aerospace Sciences Meeting, Jan. 6, 1986.

4. Michael Selz, "Commercializing Work of Federal Labs Contains Perils," *Wall Street Journal,* Oct. 30, 1991, p. B-2.

5. "The Cutting Edge: Science and Technology for America's Future," Hearings before the Subcommittee on Investigations and Oversight of the House, Science, Space, and Technology Committee, 1988, pp. 29–30.

6. "The Role of Science and Technology in Economic Competitiveness," a report for the National Science Foundation prepared by the National Governors Association Center for Policy Research and Analysis and the Conference Board, September 1987.

7. *Wall Street Journal,* Aug. 27, 1991, p. B 2.

Chapter 6

1. *Wall Street Journal,* April 23, 1991, p. A 12.

2. Britain Picks the Wrong Way to Beat the Japanese," *Science,* Vol. 252, p. 1248.

3. Susan Watts, "Esprit Gives Brussels a Good Name," *New Scientist,* Dec. 16, 1989, pp. 16–17.

4. Michael E. Porter, *The Competitive Advantage of Nations,* The Free Press, a Division of Macmillan, 1990, p. 617.

5. Robert Harley in *Stereophile* magazine, June 1991, pp. 59–65.

6. Rachel Powell, "Digitizing TV into Obsolescence," *New York Times,* Oct. 20, 1991, p. 11, Section 3.

7. Drawn in part from *Science and Technology Policy—Priorities of Governments,* by C. A. Tisdell, professor of economics at the University of Newcastle, New South Wales, Australia, Chapman and Hall, London, 1981, Chapter 4.

8. *The Science and Technology Resources of Japan: A Comparison with the United States* (NSF 88-318), National Science Foundation, Washington, D.C., 1988.

9. "Prometheus: Final Report for the Launching Phase of PRO-ART," 1990.

10. Peter Fletcher, in *Electronics,* Vol. 63, No. 9, September 1990.

11. "Advanced Television, Related Technologies, and the National Interest," National Telecommunications and Information Administration, U.S. Department of Commerce, March 1989.

12. Report of a workshop sponsored by the Institute of Electrical and Electronics Engineers, United States Activities, Technology Activities Council, February 1989.

Chapter 7

1. Akio Morita with Edwin M. Reingold and Mitsuko Shimomura, *Made in Japan: Akio Morita and Sony,* E. P. Dutton, New York, 1986.

2. Ibid., p. 82.

3. Alan K. McAdams of Cornell University, for the Technology Activities Council of the Institute of Electrical and Electronics Engineers, at hearings on high-definition television before the House Space, Science, and Technology Committee, March 27, 1989, p. 373.

4. *Yen: Japan's New Financial Empire and Its Threat to America,* Simon and Schuster, 1988, p. 102.
5. *Trading Places: How We Allowed Japan to Take the Lead,* Basic Books, New York, 1989, pp. 94–95.
6. "Science and Technology Policy for the 1980s," Organization for Economic Cooperation and Development, Paris, 1981, pp. 35–41.
7. "High-Definition Television," hearings before the House Science, Space, and Technology Committee, March 22, 1989, pp. 363–365.
8. "Report of the Defense Science Board Task Force on Defense Semiconductor Dependency," Office of the Under Secretary of Defense for Acquisition, Washington, D.C., February 1987.
9. Ibid., p. 5.
10. Pentagon briefing by Norman R. Augustine, chairman of the DSB semiconductor task force and chairman of Martin Marietta Corp., Feb. 12, 1987.
11. "Sematech: Techno-Policy in Action," *Science,* April 5, 1991, p. 23.
12. *Made in America,* a report by the MIT Commission on Industrial Productivity, Harper Perennial Edition, New York, 1990, p. 10.
13. Dr. Rodney S. Rougelot, president and chief executive officer of Evans and Sutherland Computer Corp. at hearings before the Subcommittee on Science, Research, and Technology of the House Science, Space, and Technology Committee, June 20, 1989, p. 113.
14. Lloyd M. Thorndyke, senior vice president of Control Data Corp., at a hearing on supercomputers before the Subcommittee on Science, Research, and Technology of the House Science, Space and Technology Committee, June 20, 1989, p. 85.
15. *Wall Street Journal,* Aug. 6, 1991, p. 1.

Chapter 8

1. *The Technology Pork Barrel,* Linda R. Cohen, Roger G. Noll, et al. The Brookings Institution, Washington, D.C., 1991, p. 103.
2. Ibid.
3. "Report of the Presidential Commission on the Space Shuttle Challenger Accident," Washington, D.C., 1986, pp. 164–178.
4. "Summary and Recommendations of the Advisory Committee on the Future of the U.S. Space Program." December 1990, p. 3.
5. "Report of the HST Strategy Panel: A Strategy for Recovery," Space Telescope Science Institute, Baltimore, 1991.
6. David E. Sanger, "Japan Wary as U.S. Science Comes Begging," *New York Times,* Oct. 27, 1991, Sect. 4, p. 16.

Chapter 9

1. Michael E. Davey, Christopher T. Hill, and Wendy H. Schacht, "Establishing a Department of Science and Technology: An Analysis of the Proposal of the

President's Commission on Industrial Competitiveness," Congressional Research Service of the Library of Congress, May 30, 1985.

2. Paul H. Richanbach et al., "The Future of Military R&D: Towards Flexible Acquisition Strategy," The Institute for Defense Analyses, Alexandria, VA, 1990, pp. 11–13.

3. "Global Competition: The New Reality. The Report of the President's Commission on Industrial Competitiveness," Government Printing Office, Washington, D.C., January 1985.

4. Michael L. Dertouzos, Richard K. Lester, and Robert M. Solow, *Made in America: Regaining the Productive Edge,* HarperCollins Publishers, New York, 1990, pp. 110–111.

5. "Gaining New Ground: Technology Priorities for America's Future," Council on competitiveness, Washington, D.C., 1991.

6. Andrew S. Rappaport and Shmuel Halevi, *The Computerless Computer Company,* Harvard Business Review, July-August, 1991, p. 69; replies September-October, 1991, p. 140.

Chapter 10

1. These figures may differ from earlier quoted defense R&D figures, for they include Department of Energy nuclear work, approximately $3 billion.

2. "Managing Critical Technologies: What Should the Federal Role Be?" Proceedings of a Conference Board Forum in Washington, D.C., Dec. 14, 1989.

3. "Civilian Technology and Public Policy," in *Defense Science and Public Policy,* edited by Edwin Mansfield of the Wharton School, University of Pennsylvania, W. W. Norton, New York, 1968, p. 221.

4. Robert B. Reich, "Why the U.S. Needs an Industrial Policy," *Harvard Business Review,* January/February 1982, pp. 76–81.

5. For example, Robert B. Reich, *Tales of a New America,* Times Books, a division of Random House, New York, 1987, pp. 228–232.

6. "More like Them: The Political Feasibility of Strategic Trade Policy," *Brookings Review,* Spring 1991, p. 14.

Index